Sustainable Development

Anupama Puri Mahajan

Sustainable Development

A Paradigm Shift from Development Administration

PETER LANG

Chennai · Berlin · Bruxelles · Lausanne · Oxford · New York

Bibliographic information published by the Deutsche Nationalbibliothek. The German National Library lists this publication in the German National Bibliography; detailed bibliographic data is available on the Internet at http://dnb.d-nb.de.

A catalogue record for this book is available from the British Library.

Library of Congress Cataloging-in-Publication Data
Names: Mahajan, Anupama Puri author
Title: Sustainable development : a paradigm shift from development administration / Anupama Puri Mahajan.
Description: Oxford ; New York : Peter Lang, [2025] | Includes bibliographical references and index. |
Identifiers: LCCN 2025015839 (print) | LCCN 2025015840 (ebook) | ISBN 9781803746616 paperback | ISBN 9781803746623 ebook | ISBN 9781803746630 epub
Subjects: LCSH: Sustainable development
Classification: LCC HC79.E5 M3325 2025 (print) | LCC HC79.E5 (ebook)
LC record available at https://lccn.loc.gov/2025015839
LC ebook record available at https://lccn.loc.gov/2025015840

ISBN 978-1-80374-661-6 (print)
ISBN 978-1-80374-662-3 (ePDF)
ISBN 978-1-80374-663-0 (ePub)
DOI 10.3726/b22158

Published by Peter Lang Pvt Ltd, Chennai, India
info@peterlang.com - www.peterlang.com

This publication has been peer reviewed.

I dedicate this book to the indelible memory of my mother, Late. Karuna Puri. Her profound influence, relentless encouragement, and unshakable belief in my potential ignited my path as an author. She was a beacon of strength, wisdom, and grace, and it is her enduring legacy that propels my every word and action. Her spirit continues to be the driving force behind my work, shaping this book and every endeavour I undertake.

Contents

Preface

In recent decades, the world has experienced profound changes in economic, social, and environmental spheres. Globalisation has intensified interconnectedness, making it clear that the challenges faced by one nation can have far-reaching effects on others. From climate change to migration, from poverty to political instability, the issues facing governments and institutions today demand a comprehensive, forward-thinking approach. Development administration, a field born out of the need to address the developmental challenges of newly independent nations after World War II, has evolved significantly in response to these global challenges.

The traditional models of development administration, which focused largely on economic growth and industrialisation, are no longer sufficient in a world where the consequences of environmental degradation, social inequality, and unsustainable practices are becoming increasingly apparent. In the aftermath of the Second World War, the newly decolonised nations of Africa, Asia, and Latin America were in urgent need of development strategies that could lift their populations out of poverty, build infrastructure, and create institutions of governance. During this period, development was largely seen as a linear process, with economic growth as the primary driver of progress. However, this approach soon revealed its limitations, particularly as the environmental and social costs of rapid industrialization became clear.

The theory of development administration was initially built upon Western models that were exported to the Global South, with little regard for the local political, social, and economic contexts. Scholars and policymakers assumed that what had worked in Europe and North America could be easily replicated in countries with vastly different histories, resources, and governance structures. This assumption proved to be deeply flawed, as many of these newly independent nations struggled to implement development policies that were ill-suited to their specific needs.

Moreover, the exclusive focus on economic growth often ignored critical issues of social justice and environmental sustainability. Industrialization and urbanization, while driving economic progress, frequently led to environmental degradation and the marginalization of vulnerable populations. The extraction of natural resources, deforestation, and the expansion of agricultural lands without proper environmental oversight resulted in loss of biodiversity and disrupted ecosystems, pushing many societies toward unsustainable paths. This became increasingly evident as the global environmental crisis deepened in the latter half of the twentieth century.

A new vision for development administration began to take shape, one that recognised the interconnectedness of economic, social, and environmental well-being. The publication of the Brundtland Commission's report *Our Common Future* in 1987 marked a pivotal moment in the global discourse on development. The concept of 'sustainable development' articulated in the report highlighted the need for development policies that meet the needs of the present without compromising the ability of future generations to meet their own needs. This shift in thinking introduced a holistic approach to governance and development, one that emphasises the integration of environmental stewardship, social equity, and economic growth.

Sustainable development emerged as a response to the failures of the earlier models. It represented not just a change in development theory but a paradigm shift in how governments, institutions, and international organisations approached the challenges of the twenty-first century. At the heart of this shift is the recognition that economic growth alone is not enough. It must be coupled with efforts to protect the environment, reduce inequality, and promote social inclusion. These principles have been enshrined in the United Nations' Millennium Development Goals (MDGs) and, more recently, the Sustainable Development Goals (SDGs). Both have set ambitious targets for eradicating poverty, promoting gender equality, ensuring access to clean water and sanitation, combating climate change, and fostering peaceful, inclusive societies.

The adoption of the SDGs by 193 member states of the United Nations in 2015 marked a significant global commitment to addressing the complex and interrelated challenges facing humanity. The SDGs represent a comprehensive agenda for sustainable development, covering a wide range of issues from economic growth to environmental protection, from health and education to peace and justice. Unlike the MDGs, which focused primarily on developing countries, the SDGs are universal, recognising that every nation, regardless of its level of development, has a role to play in achieving sustainability.

The paradigm shift towards sustainable development has required public administrators and policymakers to rethink how they approach governance. In many countries, this has meant reimagining the role of government, moving away from traditional top-down approaches to development and towards more participatory, inclusive models of governance. This shift has also placed a greater emphasis on the need for international cooperation, as many of the challenges we face today — climate change, pandemics, transnational migration, and the like — cannot be addressed by any single country acting alone.

International organisations like the United Nations, the World Bank, the International Monetary Fund (IMF), and the Organisation for Economic Co-operation and Development (OECD) have played pivotal roles in promoting sustainable development around the world. These organisations provide not only financial and technical assistance but also serve as platforms for dialogue and cooperation between nations. The Paris Agreement on climate change, adopted in 2015, is one example of how international cooperation has been instrumental in addressing global challenges. By bringing together governments, businesses, and civil society, the agreement has set ambitious targets for reducing greenhouse gas emissions and limiting global warming to well below two degrees Celsius.

Despite these efforts, the implementation of sustainable development policies remains uneven across the globe. Many countries, particularly in the Global South, continue to face significant challenges in achieving the SDGs. Financial constraints, political instability, weak institutions, and corruption are just some of the obstacles that have hindered progress. In some cases, governments have been unable to mobilise the resources needed to implement sustainable development programmes, while in others, entrenched interests have resisted the changes necessary to achieve more inclusive and environmentally sustainable growth.

This book aims to address these challenges by offering a comprehensive analysis of development administration and its evolving relationship with sustainable development. It examines the theoretical foundations of development administration, tracing its evolution from its origins in the post – World War II era to its current focus on sustainability and inclusivity. The book also explores the various models of sustainable development that have emerged in different parts of the world, highlighting both successes and failures in the implementation of these models.

One of the key themes of this book is the importance of citizen-centric governance in sustainable development. Around the world, citizens are increasingly demanding greater accountability, transparency, and participation in decision-making processes. This shift towards more participatory forms of governance reflects a broader recognition that development cannot be achieved without the active involvement of citizens. Whether through digital platforms, grassroots movements, or direct policy engagement, citizens are playing an increasingly important role in shaping the policies that affect their lives. In many countries, this has led to the development of new governance models that prioritise citizen participation and collaboration with civil society organisations.

The role of gender in sustainable development is another critical issue discussed in this book. Gender inequality remains one of the most pervasive forms of social injustice around the world, and it is a major barrier to achieving sustainable development. Women and girls often face systemic barriers that limit their access to education, healthcare, and economic opportunities, and they are disproportionately affected by poverty and environmental degradation. SDG 5, which focuses on achieving gender equality, is not only a goal in itself, but also a prerequisite for achieving many of the other SDGs. This book explores the ways in which gender equality can be promoted through sustainable development policies and how empowering women and girls can lead to more inclusive and resilient societies.

Environmental stewardship is another central theme of this book. As the global environmental crisis deepens, the need for strong governance structures that can effectively manage natural resources and protect ecosystems has become increasingly urgent. Many of the challenges we face today, such as climate change, deforestation, and biodiversity loss, are the result of unsustainable practices that have persisted for decades. Dealing with these challenges will require not only technical solutions but also political will and ethical leadership. This book examines the role of governance in promoting environmental sustainability, highlighting examples of successful environmental policies and initiatives from around the world.

Ethical governance is indispensable for sustainable development. Corruption remains a significant challenge in many parts of the world, undermining public trust and eroding the effectiveness of institutions. The international community has made strides in addressing corruption through initiatives like the United Nations Convention Against Corruption (UNCAC), yet the challenges remain immense. Without strong governance structures that emphasise transparency, accountability, and ethical leadership, the path to sustainable development will remain fraught with obstacles. The institutional frameworks and ethical codes that have been established to curtail corruption and promote good governance, offering insights into how these frameworks can be strengthened.

The structure of this book is designed to provide readers with a holistic understanding of both traditional and contemporary issues in development administration. The first chapters examine the foundational theories of development administration and their relevance in today's globalised world. The subsequent chapters explore the rise of sustainable development as a paradigm, offering real-world case studies from different countries to highlight the ways in which sustainable development is being implemented on the ground. Each chapter studies a critical dimension of sustainable development, with a focus on how these elements intersect to form a comprehensive framework for governance. Whether examining the role of international organisations, the importance of citizen participation, or the need for environmental stewardship, this book provides a thorough exploration of the issues at the heart of sustainable development.

This book aims to contribute to the ongoing global effort to promote sustainable development by providing readers with the tools and knowledge necessary to navigate the complexities of the field. It is my hope that this work will inspire thoughtful reflection and actionable solutions for a more sustainable, just, and inclusive world.

Figures

Tables

Development Administration

This chapter provides a comprehensive overview of development administration, focusing on its evolution and significance in governance. It begins by exploring the emergence and definition of the term, tracing its origins to post-World War II efforts to support development in newly independent nations. It contrasts development administration with traditional administration, emphasising the former's focus on economic growth and social change and the latter's focus on routine tasks and stability, whereas. The chapter then delves into the nature and context of development administration. It recognises the unique challenges developing countries face, such as poverty and political instability, which require dynamic and flexible administrative approaches. It discusses the scope and features of development administration, highlighting its role in planning, coordinating and implementing policies for national development. This chapter additionally examines administrative patterns in developing countries, showing how governments adapt structures and processes to meet developmental goals. It presents various approaches to development administration, emphasising different strategies to achieve growth and modernization. Finally, the chapter evaluates the present status of development administration, assessing its successes, challenges and ongoing relevance in today's globalised world.

Emergence of development administration

There is evidence that public administration has long been a part of human civilisation, giving governance – in modern terminology – to all citizens under its realm, whether a kingdom or a republic. The primary aim of public administration is to cater to the development of undeveloped areas for the upliftment of the vulnerable sections of society. 'Development', 'modernization' and 'modernity' are related concepts that refer to major societal transformations involving complex social, economic and political changes. Development generally occurs 'when an index of what is deemed desirable and relatively preferable increases in magnitude'.[1]

1 Ralph Braibanti and Joseph J. Spengler, *Tradition, Values and Socio-economic Development* (Durham, NC: Duke University Press, 1961).

Ferrel Heady and his colleagues chose the term 'developing' to describe modern nations as a group with diverse circumstances. Riggs pointed out that nations move historically from 'agraria' to 'industria'.[2] In developing countries, development has two primary goals: nation-building and progress. Milton J. Esman describes nation-building as 'the deliberate fashioning of an integrating political community within fixed geographic boundaries in which the nation-state is the dominant political institution'.[3] Development administration is integral to resolving the challenges and promoting growth in socioeconomically marginalised and backwards areas.

Development administration emerged in the post-Second World War era. Many countries gained their freedom from imperialism and became separate nations. These newly formed nations were free from the colonial powers. Still, freedom has been questioned by scholars like Amartya Sen. He posed the question of whether freedom was attained in totality in equality. He stated that development and freedom are two inseparable parts of independence.

The circumstances under which development administration emerged as an independent discipline can be understood from the following situations and efforts. The governments of the new nations realised that the colonial rule administrative systems would not be able to cater to their developmental needs. Most of the administrators serving in the colonial administration opted to mercernarise their services for whichever nation wanted to hire their administrative services for a higher price.

The 1950s

The Marshall Plan was introduced in the post-Second World War period to recover the development of Western Europe's economies. The plan was based on the principles of self-help and mutual help involving whole populations. Its significant contribution was to produce the concept of the comprehensive country programme in economic and social development. The Marshall Plan inspired low-income countries of the Global South to follow the example. The planning for development gained momentum, which the United Nations and international private agencies such as the Ford Foundation and The Rockefeller

2 Fred W. Riggs, 'Agraria and Industria—Toward a Typology of Comparative Administration', in *Toward the Comparative Study of Public Administration*, ed. by William J. Siffin (Bloomington, IN: Indiana University Press, 1957), 49–78.

3 Milton D. Esman, 'The Politics of Development Administration', in *Approaches to Development, Politics, Administration and Change*, ed. by John D. Montgomery and W.J. Siffin (New York: McGraw-Hill, 1966), 69–70.

Foundation promoted in the 1960s. The International Bank for Reconstruction and Development (IBRD) was established to develop a new international monetary system in the 1950s.

The 1960s

In the 1960s, Walt Whitman Rostow proposed the five stages of the economic growth model, but the world was looking for a comprehensive development plan for inclusive development. He gave an approach with an historical perspective that specified five stages to reach the economically developed stage, which are: traditional society, preconditions for take-off, growth, drive to maturity and age of mass consumption. However, this theory led to economic development without inclusivity. Underdeveloped countries could not benefit from this concept. The world moved towards a search for alternative development models and strategies.

Significant developments in public administration towards development administration

The role of public administration gained importance, and the following two significant efforts emerged: the Committee on Comparative Politics of the Social Science Research Council (SSRC) and the Comparative Administration Group of the American Society for Public Administration (ASPA).

The Social Science Research Council's (SSRC) Committee on Comparative Politics

The SSRC's Committee on Comparative Politics was a political development group that emphasised the role and functions of the bureaucracy in maintaining and stabilising political order. Gabriel Almond and Lucian Pye studied the politics of 'non-Western' states. Pye researched the non-Western political processes, while Almond studied the politics of the issues of the developing countries, comparative politics and the political culture of the developed and developing countries. The Ford Foundation studied theoretical issues and their implications for public policy. They contended that liberal democracy could be quickly established in developing countries and saw mass participation as a way to their development.

The American Society for Public Administration's (ASPA) Comparative Administration Group

The ASPA's Comparative Administration Group played a pivotal role in shaping the study of development administration. Ferrel Heady's work, *A Comparative*

Perspective, became a seminal text that contributed to the understanding of public administration in developing nations. The Comparative Administration Group and ASPA's contributions emphasised the need for strong administrative systems to facilitate nation-building in newly independent countries.

Nation-building, emerging as a reaction against colonialism, was seen as a critical process for countries striving to free themselves from the remnants of colonial rule. The ideology of 'development' in these nations focused on creating robust governmental structures and leadership. The belief was that such institutional strength would act as a foundation for achieving socio-economic progress. Developing countries, inspired by the paths taken by developed nations, sought to establish effective governance to drive both political stability and economic independence.

However, socioeconomic progress, particularly social change, did not always receive equal emphasis in this phase. While governments aimed to boost economic independence and improve the standard of living, social dimensions such as equity, education and health were often sidelined. The assumption was that following Western models of economic growth would automatically lead to social improvements. In reality, a strong government that was formed during the nation-building process needed to actively address both economic and social aspects for balanced, sustainable progress. This interdependence highlights that nation-building and socioeconomic progresses must go together, with public administration playing a crucial role in shaping and sustaining both.

Development administration defined

There is no clear-cut definition of development administration, as it is more of an abstract concept that needs to be described. Many scholars have attempted to define it. Some definitions have been given by eminent scholars including George F. Gant, Edward Weidner, Saul M. Katz, F.W. Riggs, David C. Stone, and Amartya Sen. What follows is a closer look at each scholar's perspective.

George F. Gant is known as the father of development administration. He first described 'development' as not an absolute but a relative condition, with no country qualifying as fully developed. The term' development administration' denotes 'the complex of agencies, management systems and processes a government establishes to achieve its development goals. Development administration is the administration of policies, programs and projects to serve development and purposes'. It is characterised by its purposes, which are 'to stimulate and facilitate defined programs of social and economic progress' in two ways: by its loyalties, which are to the public rather than to vested interests, and by its attitudes, which are 'positive rather than negative, persuasive rather than restrictive'.[4]

Edward Weidner defined development administration as the process of guiding an organisation towards the achievement of development objectives. As per his definition, it is action-oriented and places administration at the centre in facilitating the attainment of development objectives.[5] For Saul M. Katz, development is a major societal change from one state of national being to another, more valued state. Development planning involves the rational direction and acceleration of this societal change towards the specified objectives. Development action is an essential part of development planning as a process.

F. W. Riggs defined development as 'a process of increasing autonomy of social systems, made possible by rising level of diffraction'. Diffraction is considered necessary by Riggs as a sufficient condition for development.[6] He added that development administration refers to administrative systems or organisations centrally concerned with achieving progressive socio-economic and political goals and innovating attitudes and operations. For David C. Stone, development is a continuous process of formulating, reformulating and implementing interrelated decisions to realise stated development ends in a prescribed time sequence and by optimum means. For Amartya Sen, lastly, development can be seen in two senses. One is broader, and the other is narrower. In his opinion, development has a wide meaning: the expansion of absolute freedom. Through development, people can have opportunities to expand their freedom.

Traditional public administration and development administration

Development administration deals with achieving various developmental goals and bringing about economic, political, cultural and social changes in a developing society, but it differs from traditional public administration. The differences are given in Table 1.

4 George F. Gant, 'The Concept of Development Administration', in *Comparative Public Administration: The Essential Readings*, ed. by Eric E. Otenyo, Nancy S. Lind and Lawrence R. Jones (Oxford: JAI Press, 2006), 45–68.

5 Edward W. Weidner, *Technical Assistance in Public Administration Overseas: The Case for Development Administration* (Chicago: Public Administration Service, 1964).

6 *Frontiers of Development Administration*, ed. by Fred W. Riggs (Durham, NC: Duke University Press, 1970).

Table 1 Comparison between traditional public administration and development administration

No.	Traditional public administration	Development administration
1	Open loop variety with weak control feedback and lacks purposeful development action, which is hard to achieve.	Self-regulatory features of closed-loop control with clear and strong feedback circuits.
2	Not responsive to social and economic development demands, legislatures and people.	Persuasive and positive response when conducting business with other countries.
3	Not concerned with rectifying inequities in the social system.	Explicitly concerned with removing all types of inequities in the social system.
4	Did not encourage economic growth or equitable distribution of benefits in the society.	Promotes innovation and change, as well as quality of administration of justice for all.
5	Performed functions only of law and order, building roads and settling disputes.	Performs developmental functions.
6	Internal administration includes efficiency in their specified functions, achievement of economies in time, personnel and materials, and management of an organisation's processes and methods.	External administration refers to the activities and processes required to achieve the agency's purpose outside the agency, such as developing relations with other agencies and having client participation.

The nature and context of development administration

Development administration refers to organising and managing resources to achieve socio-economic growth, modernization and social transformation, particularly in developing countries. Its nature is dynamic, goal-oriented and innovative, focusing on fostering change and development. This is unlike traditional administration, which is more concerned with routine tasks and maintaining order. Development administration emphasises adaptability, responsiveness and flexibility in the face of evolving challenges.

The context in which development administration operates is shaped by the socio-political and economic realities of developing nations. These environments often face complex issues like poverty, underdevelopment and political instability. As a result, development administration must address these challenges by promoting equitable growth, institutional capacity-building, and inclusive governance to uplift marginalised communities and foster long-term sustainability.

F. W. Riggs was one of the pioneers in building the development administration subject. He said that development administration refers to the organised efforts to carry out programmes or projects developed by those involved to serve developmental objectives. Riggs offers two points to consider, focusing on freedom and variability, while building the context of development administration. In the case of freedom, Riggs argued that an individual does not have the freedom to shape one's physical, human and cultural environment in a primitive society because society is closed and holds more power. This implies that a society must initiate collective decision-making to bring about political and administrative change.

As for variability, Riggs believed that countries vary in their capacity of decision-making to change their environment due to their varying capacity to carry the burdens of their developmental issues. Riggs proposed various stages or determinants of development administration at the political, economic and organisational levels to understand administrative development and learn what administration principles are relevant to different government systems.

Environmental systems

Riggs focused, in particular, on environmental conditions and stages of development. In the case of environmental conditions, Riggs argued that every polity has a varied environmental or ecological development level. The relationship between a polity and its environment varies between developed and developing nations. A developing country has more capability for change.

Stages of development

As for stages of development, it is generally believed that the administrative capacity and development depend on the developed countries' environmental constraints and administrative principles, but this cannot be applied to developing countries. Riggs opposed this belief and said that it can be accepted that environmental problems affect the administration of development programmes but do not affect the relevance of administrative principles.[7] The stages of development – political, economic and organisational – determine the applicability of administrative doctrines and govern the development of administrative capabilities. Hence, the stages of development affect administrative development. Riggs termed the stages of development as 'the context of development administration'.

7 Environmental or ecological conditions refer to a society's social, political and economic conditions.

The various perspectives of the 'context of development administration' given by Riggs including the political context, economic context, and social context.

The political context

Riggs argued that public bureaucracy is the primary mechanism for governments to implement development programs supported by legislative bodies, courts, political parties and citizens. While bureaucracy performs essential functions, it often overlaps with political roles. A balance between politics and administration is crucial.[8] Weak bureaucracy can hinder administrative improvement, whereas dominant bureaucracy fills cabinet positions with career officials or military appointees. Effective government structure relies on strong political bureaucracy and balanced administrative doctrines. However, Western doctrines may not suit non-Western nations, requiring a balance of social system needs for successful implementation.[9]

The economic context

Riggs argued that a balanced polity requires an adequate salary system, funded by taxation and economic production across primary, secondary and tertiary sectors. Industrialised societies can support well-paid bureaucracies due to high production levels, which help prevent corruption and political pressure, promoting effective administration and economic growth. Nonetheless, Western financial aid to non-Western countries was insufficient, leading to corruption among officials. New nations faced inefficient, growing bureaucracies that did not add societal value. Poor salary structures in countries with low economic development led to ineffective bureaucracy and political instability proving that established administrative doctrines fail in low-growth societies.

The social context

Most public administration theorists build theory on administration in a relatively advanced society. The theory of organisation is practical in a developed society as it manipulates the environment to its advantage. On the contrary, building institutions in a less developed society is complex. It can happen only if a society has reached the 'take-off' stage of economic development, as given by Walt Rostow. Development picks up only after a society reaches the 'take-off' stage, with organised activities and administrative doctrines finding a relevant place.

---

8 Fred W. Riggs, 'Bureaucracy', in *Frontiers of Development Administration*, ed. by Fred W. Riggs (Durham, NC: Duke University Press, 1970), 89–106.

9 Extra-bureaucratic institutions include elected representatives or the constitutional system.

The three given contexts, in sum, are interlinked as the prevalence of one of these three conditions can bolster the rest of the two. A balanced polity, organisational maturity and an adequate salary system of bureaucracy determine a society's development level.

The scope of development administration

The concept of development is elusive, encompassing life conditions, goal attainment, capacity growth and change. Evaluating a society as underdeveloped, developing, or developed is challenging due to changing indices of essential services. Fifty years ago, food, shelter and clothing were fundamental. Today, health, education and internet access are included. People play a crucial role in development, utilizing natural resources. Development programs target people as both beneficiaries and instruments of progress. Some significant areas where development administration reaches out to people are discussed here (see Figure 1).

1. Poverty reduction: George F. Gant emphasised reducing poverty as the primary goal of development administration, which includes addressing hunger, housing, clothing, and health. The World Bank stated that about 40 per cent of the population in developing countries, 800 million people,

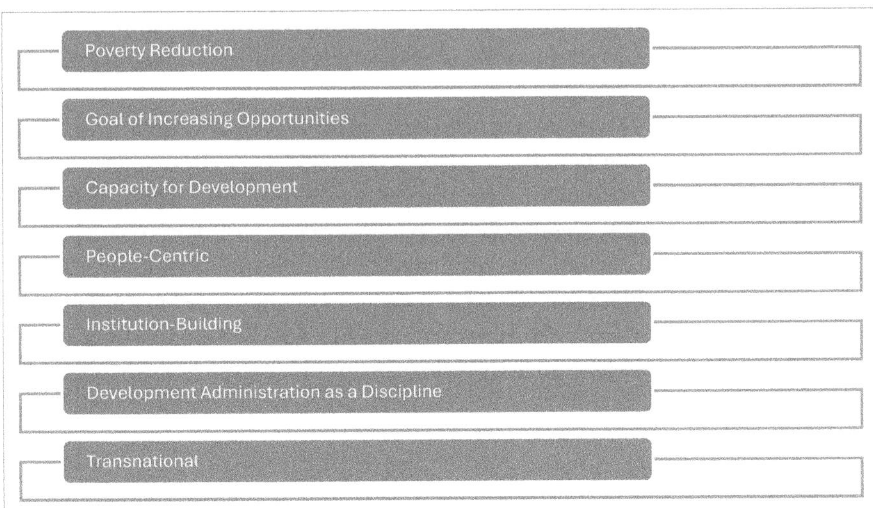

Figure 1 The scope of development administration

live in absolute poverty.[10] Per capita income often misrepresents regional disparities. In 1999, Amartya Sen suggested understanding poverty as a lack of essential abilities rather than just focusing on income, emphasising that low income often leads to limited opportunities and capabilities.[11] Therefore, development administration's foremost goal is to reduce or eliminate poverty, ensuring sustainable living for all.

2. The goal of increasing opportunities: The scope of development administration extends beyond reducing poverty. Development must also consider quality of life, dignity, human rights, and liberty. People need ample choices and opportunities because bare subsistence is insufficient. Amartya Sen states, 'Expansion of freedom is both the primary end and the principal means of development', emphasising that development involves removing unfreedoms that limit choices and opportunities for exercising reasoning agency'. [12]
3. Capacity for development: Development administration strives to increase the capacity for development as a goal. Capacity for development refers to the ability of a nation or community to effectively plan, implement, and sustain programs aimed at social, economic and environmental progress. It involves building institutions, fostering strong leadership, and ensuring resource availability. Development capacity also includes human resources, developing skills, knowledge and innovation among the population. It requires robust governance structures that promote transparency, accountability and inclusiveness. A high capacity for development allows countries to overcome challenges like poverty, inequality and environmental degradation, and to sustain long-term growth and improvements in quality of life. Continuous investment in education, technology and infrastructure strengthens this capacity.
4. People-centric: Development administration is people-centric and focuses on the ultimate beneficiary, the populace. The nation-state must be strengthened socially, economically, politically and culturally, acting as an instrument to implement policies that ensure contentment to the people. People, as the most valuable resource, drive economic development. The state must formulate demographic policies to maintain an optimal population size, preventing it from becoming a burden. Development administration should present an attractive framework to motivate participation,

10 World Bank, *World Development Report* (New York: Oxford University Press, 1978).
11 Amartya Sen, *Development as Freedom* (Oxford: Oxford University Press, 1999).
12 Ibid.

offering better economic well-being and broader work and leisure opportunities for current and future generations while raising awareness of available opportunities.[13]

5. Institution-building: New developing countries often lack the institutions necessary to implement development projects effectively. An institutional framework is essential to transform development policies and projects into action, encompassing departments, schools, hospitals and budgeting agencies. Institutions stabilise economic and social behaviour patterns. George F. Gant defined an institution as a system of actions with key qualities:
   - Effective performance of intended functions
   - Acceptance by society and the local environment
   - Representation of individuals' and groups' needs and relationships
   - Value for the people it serves
   - Strength and flexibility for survival, supported by financial, personal and political resources

   Institutions must possess the capability, public acceptance and survival capacity. Otherwise, they should be reformed or discarded. Flexibility is crucial for maintaining effectiveness and responsiveness. Developing nations require institutions and reforms to drive societal development. Modernised institutions are central to implementing development plans, but need skilled, trained personnel. Enhancing educational capacities ensures good staff and effective leaders provide control with legal and administrative authority. Institutions must also include two elements. First, they require assurance of financial resources for continuous operations through appropriations, revenues, borrowings and the like. Second, the require administration and organisational hierarchy with a suitable job structure and harmonious environment for wise decision-making. Effective development administration also relies on well-coordinated institutional components.
6. Development administration as a discipline: Since the 1950s, there has been growing recognition of the significance of development administration. As a result, numerous organisations and scholars began to focus on exploring and shaping the field, contributing to its theoretical foundation and practical application. Universities and colleges have included the course of

13 George F. Gant, 'The Concept of Development Administration', in *Comparative Public Administration*, ed. by Eric E. Otenyo and Nancy S. Lind (New York, NY: Elsevier, 2006), 45–68.

development administration in their curriculum, which has helped new administrators understand the complexities involved in development.
7. Transnational: Development administration does not limit itself to the national level and goes beyond boundaries. New nations may evolve their developmental strategy relating to planning but have to depend on the West for technical and financial aid. The developing countries depend on other countries to develop trade and business relations and to get their technical knowledge. Development models are built on Western models with modifications to suit their needs.

Therefore, the scope of development administration encompasses both the formulation and implementation of policies aimed at achieving social and economic progress. It focuses on improving governance structures, enhancing administrative capacity and addressing the needs of marginalised communities. Ultimately, development administration seeks to foster sustainable growth and equitable development.

Features of development administration

After examining the scope of development administration, its key features are outlined and illustrated in Figure 2.

1. Dynamic: Development administration simply means goal-oriented development. Development can occur in a dynamic society affected by its

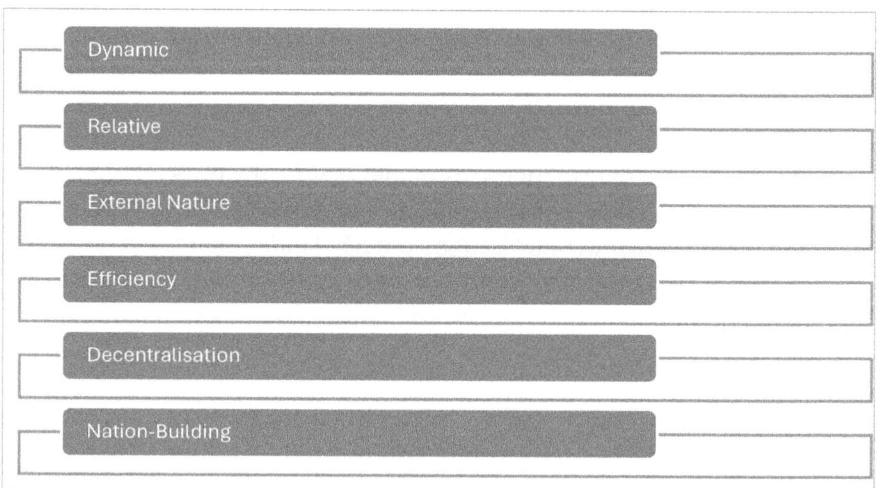

Figure 2 Features of development administration

ecology, that is, its internal and external environment. Change orientation is an essential characteristic that will bring about development. Contrarily, societies that maintain a status quo and do not wish for disruption cannot foster development. Development administration is change-oriented in a dynamic society.

2. Relative: Development administration is dynamic because societies continuously evolve and the nature of essential services change. Development is relative and ever-changing, never absolute. Even developed nations like the United States have areas for improvement. This relativity creates fluidity in development administration, extending to societal desires and expectations. As people gain information and awareness, their expectations increase, driving them to attain new goals and improve their standard and style of living.
3. External nature: Development administration deals with those activities and processes required to establish interagency collaboration outside the control of a state agency to implement policies and programmes. It also involves client participation. Interagency collaboration is of two types: informal and formal. Informal collaboration refers to unstructured relationships that are voiced in conferences and so on. Formal collaboration includes contracts or agreements for systematic cooperation. There must be clearly defined patterns of client participation, such as cooperatives, farmers' organisations and trade associations. They can be formal or informal.
4. Efficiency: Development administration works consistently focus on efficiency. It uses the application of modern technology in management functions, especially at mid-level management, to increase effectiveness. Development administration employs measurement techniques to evaluate efficiency, which is quite direct to achieve the developmental goals. Simply put, if the administration's goal is achieved, then it is good administration, and if not, then it is terrible. Since the purpose of development administration is to achieve goals and deliver results, good or bad results are assessed in delivery. So, development administration begins with a goal and is followed by the methodology to achieve it efficiently.
5. Decentralization: Development administration favours decentralization to diffuse the power and business at the centre via delegation of authority to rural agencies. The planning system, information and reporting systems, human resource development and coordination are aimed at decentralization. Hence, development administration is not only the administration of development, but also the development of administration. It enhances the bureaucratic capacity to reinforce development by coordinating the work of all agencies in decentralization.

6. Nation-building: Since the development administration emerged after the formation of new nations in the post-Second World War period, initially it had to deal with nation-building though it is a continuous process. Esman describes nation-building as 'the deliberate fashioning of an integrated political community within fixed geographic boundaries in which the nation-state is the dominant political institution'.[14] Specialised and technical staff with expertise are essential to development administration to carry out the developmental tasks successfully.

The ministries responsible for nation-building require specialised staff to carry out operations in the field. Specialists must be inducted into the civil services so that they understand the developmental needs and think of innovative ways to execute developmental projects. Nation-building should also include devising new tools and mechanisms by setting up agencies for planning at the state and district levels. Thus, development administration deals with formulating new agencies and reorientating existing ones to take up socio-economic development's arduous and complex functions. It must involve the government, the private sector and the people to succeed.

Administrative Patterns in Developing Countries

Developing nations typically follow specific administrative patterns, with strong governance essential to driving development. It has been accepted that administrative capability and machinery are inadequate for development. Some administrative patterns found in the administration of developing countries are depicted in Figure 3 and described as follows.

1. Imitative instead of indigenous: Developing nations often adopted Western administrative structures due to colonial influence, mirroring former colonial powers.[15] British administration, considered the best, was authoritarian and paternalistic, reflecting their bureaucratic style.
2. Lack of administrative skills in developmental programmes: There is a shortage of trained administrators with management, development and technical skills amid many unskilled labour forces. Many university

14 Milton J. Esman, 'The Politics of Development Administration', in *Approaches to Development: Politics, Administration and Change*, ed. by John D. Montgomery and William J. Siffin (New York: McGraw-Hill, 1966), 85–104.

15 Donald J. Kingsley, 'Bureaucracy and Political Development with Particular Reference to Nigeria', in *Bureaucracy and Political Development*, ed. by Joseph LaPalombara (Boston: Princeton Legacy Library, 2015), 112–130.

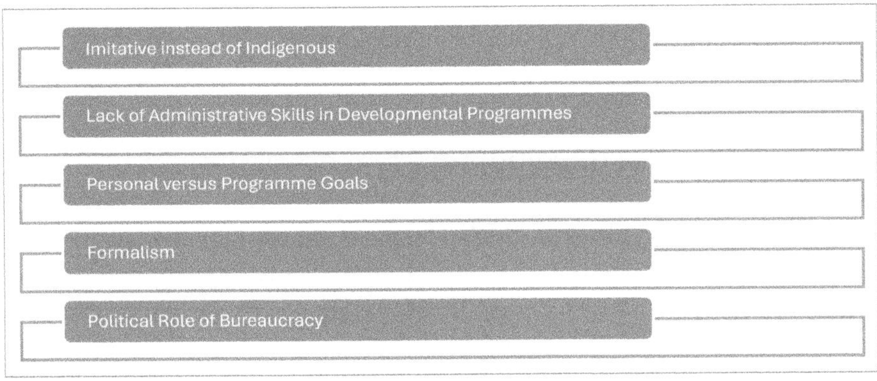

Figure 3 Administrative patterns in developing countries

graduates in developing countries remain unemployed due to unsuitable degrees or marginal institutions.[16] Training in development skills is needed to bridge the gap in filling administrative posts, leading to a time lag in building local administrative resources. Reservation policies for minority groups in developing countries also create restrictive shortcuts for administrative capacity building.

3. Personal versus programme goals: Bureaucrats often prioritise their personal agendas over productivity targets. Riggs observed this tendency, particularly in transitional societies, where traditional behaviours persist among administrative staff. Their focus leans towards status maintenance rather than program accomplishment. Despite merit-based recruitment, officials often favour loyalists, creating favourable conditions for non-merit promotions, assignments and dismissals. Another factor is bribery within bureaucracy undermines fair program execution. Increased bureaucratic size aimed at reducing unemployment can decrease efficiency.
4. Formalism: Riggs stated that there is a widespread discrepancy between form and reality, termed 'formalism'. In the transitional societies of developing countries, there is a gap between expectations and realities due to the flouting of regulations and laws, leading to the underachievement of programme targets.
5. Political role of bureaucracy: Developing countries cannot shake off the hangover of colonialism in which the bureaucracy had operational autonomy. The bureaucracy has the technical expertise but unwilling to share leading to monopolization of the institutions.

---

16 Ferrel Heady, *Frontiers of Development Administration* (Durham, NC: Duke University Press, 1970).

This analysis of the administrative patterns of the developing countries by Ferrel Heady and his colleagues led to a paradigm shift towards a separate discipline, namely development administration.

Administration of development and development of administration

Riggs identified two critical roles in development administration: implementing programs to reshape physical, human and cultural environments and enhancing governmental capacity for such initiatives. He noted that while developed nations possess robust administrative theories, these often prove unsuitable for developing countries, which require contextually relevant approaches for successful project implementation. Riggs emphasised the distinction between administering development and developing administration itself. He highlighted the challenge of applying modern bureaucratic norms to ancient systems and argued that complex bureaucracies in developing nations necessitate updated structures. In the 1960s, Stone noted that many new countries, inspired by India, were adopting development plans that combined economic and social goals, reflecting changing administrative practices in the Third World. A discussion regarding the perspectives engaged in this concept of administering development and developing administration is presented here.

Strategy for development

Development strategy and political leadership

Every country must develop a comprehensive strategy for progress articulated through its legislative framework and development plans. Stone identified several critical factors for effective development administration, focusing on political leadership. A stable and supportive political environment is crucial for successful development administration. Significant aspects of political leadership and action include the following:

1. Political action: Effective development work requires active political engagement within a robust political framework. This is especially important in societies resistant to change and entrenched in traditional practices.
2. Social Action and change: Development efforts are incomplete without incorporating social action and change. Strong, dynamic political leadership, active public participation and a well-organised administrative structure can enhance these efforts. Maximising democratic processes is essential.
3. Innovative approaches: Civil servants who use innovative and resourceful approaches will likely receive more support from political leaders. This support is crucial for demonstrating the effectiveness of developmental

plans and projects, contrasting with more rigid ideologies such as communism. Broader participation in decision-making processes can lead to more comprehensive applications.

4. Individualistic patterns: Emerging nations should be free to develop their unique democratic models and patterns, reflecting their specific needs and contexts.

Features of low-income countries

Low-income countries have the following features concerning political action to achieve development.

1. Externalities: In emerging nations, there has to be some kind of exercise of coercion to do away with external interests which hamper social progress. Political strength has to be accompanied by some exercise of power to bring about social change.
2. The leadership of minorities: Leadership in developing countries generally reflects the interests of minorities. Such societies have the most illiterate populations, and the leadership will emerge from the few with the most capabilities. Political freedom must be maintained to suppress special privileges and parasitical groups.
3. Mass appeal: A political and philosophical doctrine must be developed in a rationalistic manner having mass appeal.
4. Citizen participation: Political and administrative organisations with mass identification and citizen participation must be constituted and strengthened for developmental progress.
5. Capacity of political leadership: The capacity of political leadership and the political system must be strong enough to accommodate change. This is essential to avoid the rigidity of political party structures.

To sum up, the absence of strong political leadership leads to the use of excessive military power which deters development. All these factors must be considered while formulating a comprehensive policy for development administration.

Budgeting and planning

There is an urgent need for linking planning to financial realities for budgeted allocations. The capital current budgets must cover all available resources for a development plan. The resources included in a development plan are:

- Available revenues in the countries appropriated for ordinary government purposes

- General revenues for development programmes
- Funds raised from internal borrowing for development goals
- Funds available through financial institutions for development purposes
- Earnings from public sector undertakings (PSUs) for development
- External loans and grants

All the resources mentioned above must be passed through the budgeting process for developmental programmes. It is reasonable to allow funds for purposes other than development for industry to grow.

Administrative obstacles

Stone conducted research across twelve countries to understand administrative obstacles. The study questioned the capacity of administrative officials, and he made the following concluding observations.

1. Comprehension: The ministries do not comprehend how a development plan can help them and reject it.
2. Core plans: The ministers cannot develop and implement core plans without incurring time and cost overruns.
3. Vague projects: The developmental projects formulated by ministries are vague and no attention is paid to the feasibility, economic viability, engineering potentiality or administrative capacities required to carry out the projects.
4. Lack of coordination: There is a lack of coordination between the ministries and institutions regarding several factors that can help execute a project successfully, for example, loans, funds, raw materials, machinery and management skills.
5. Lack of coordination among departments: The departments dealing with the development work lack coordination in major social or economic development decisions.
6. Lack of governance: Local governments are not involved in the execution of developmental plans, which leads to mismanagement of progress; for example, the public works department builds roads in one season and before the construction is over, the electricity department decides to change or lay poles and break open the newly constructed roads.
7. Overload of government agencies: In developing countries, many government agencies have cropped up to add to the already confused administration task.

Problems in the administration of development by Riggs

Environmental conditions create problems for social systems, which leads to problems in the administration of development. The administrative doctrines applied to an underdeveloped or a developing society must be according to the level of development that society has reached. According to Riggs, some problems that arise in the administration of development include those related to the physical environment and human environment.

Physical environment

The geographical location of a country presents its peculiar problems, like constructing a road in a snowbound area would be more complicated than it would be in an area where it is always summer. However, governments with developed administrative capabilities can deal with both types of terrain in contrast to governments with limited capabilities. A harsh physical environment determines the degree of development of a social system.

Human environment

The human environment encompasses various aspects:

1. Demographic: In traditional societies, factors such as population size, distribution, mobility and growth rates significantly influence the effectiveness of a polity. However, in developed states, a wide variation in these factors is manageable within limits.
2. Biological: Development is hindered by a lack of suitably skilled individuals needed for an industrial society. Appropriate educational facilities can help develop human resources. Training programs bring personnel from least developed countries (LDCs) to developed nations for training, but this can lead to brain drain, where trained individuals do not return home. Building training capacity within developing countries can mitigate this issue.
3. Cultural: Cultural changes often accompany administrative reforms, posing challenges for development administration but not for the development of the administration itself. Developed social systems can better transform cultural environments to meet their needs by reshaping administrative programs to achieve cultural changes. Thus, the cultural environment can present challenges to development administration.
4. Aid: To adapt to changes in the polity, enhancing administrative capability is essential. Securing foreign technical and financial assistance can facilitate environmental transformations rather than system transformations.

To sum up, the study of development administration requires a new set of doctrines to help polities with low administrative capacity to develop the physical, human and cultural environments.

Steps to create administrative capabilities

Stone suggested some steps that can be taken to create administrative capabilities in developing countries like India. They are discussed further (see Figure 4).

1. Identification of a problem: When a problem is identified, a solution is easy to find. If the development plan is considered an administrative problem rather than an economic one, many obstacles would be removed. Administrative capacity-building is a primary step in plan implementation.
2. Central coordination: The planning agency must exercise central coordination and direction to plan administration. It must be supported by the budget, operations and maintenance office, along with team support.

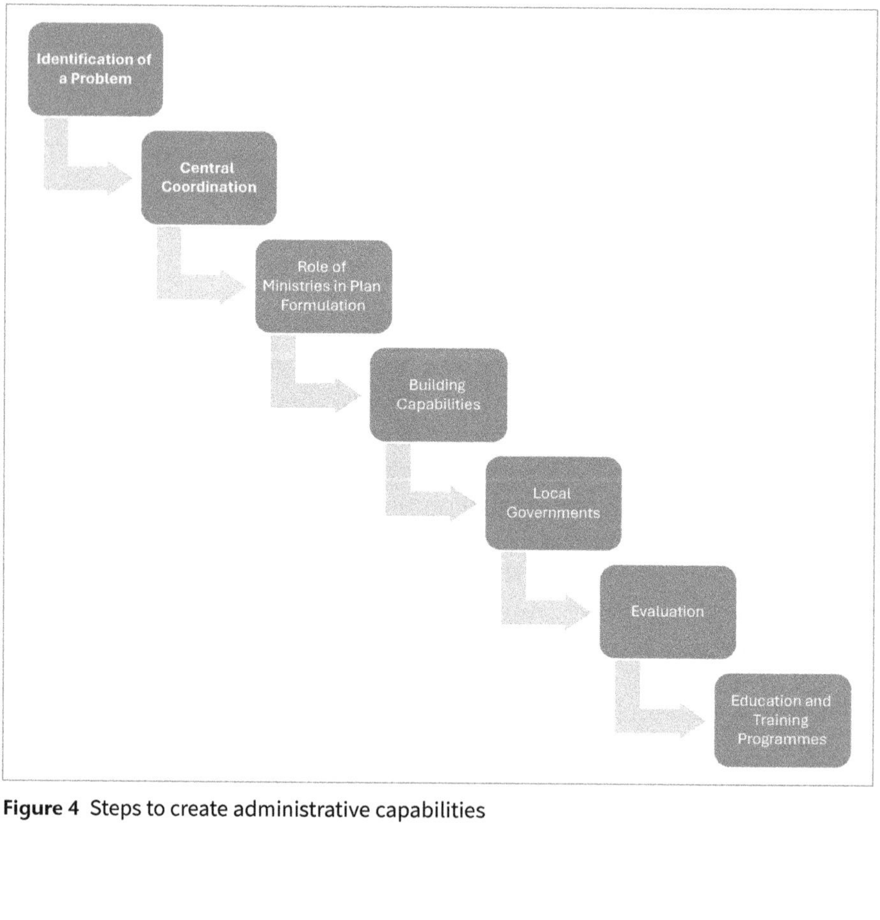

Figure 4 Steps to create administrative capabilities

3. Role of ministries in plan formulation: The ministries must be involved a lot more in the responsibility of plan formulation. Their participation would ensure more capacity in the execution of the plan.
4. Building capabilities: The central leadership must build capacities to formulate and manage projects with all the essential elements of administration.
5. Local governments: The local governments must be involved in plan preparation and execution. They must be brought into the administration fold so the development programmes can be executed pragmatically.
6. Evaluation: A practical assessment is vital to monitor the progress of the development projects. The evaluating mechanism must ensure that the development projects are progressing as per the time schedules and if not, corrective measures must be taken.
7. Education and training programmes: A severe lack of specialists and technocrats was felt during the plan implementation. This can be improved by imparting education and training to pre-service and in-service officials.

Comparative Administration Group (CAG)

The CAG suggested that the focus of the development administration study should be on the two major themes. The first is the development of government administrative institutions and capabilities, emphasising the structure and behaviour of bureaucracy. The second is the intersection of administration, policy formulation and implementation of developmental plans and programmes.

Dimensions of development administration

Milton J. Esman, a prominent scholar in the field of development administration, identified several critical dimensions or issues concerning development administration, which were categorised under three major clusters of problems. These clusters were articulated through studies and discussions held under the aegis of the Comparative Administrative Group (CAG), a group of scholars exploring the role of administration in development.[17] The three clusters are:

1. Institutional development: This cluster focuses on the establishment and strengthening of institutions necessary for development. It highlights the need to create or reform government structures, legal frameworks and administrative systems that effectively support development initiatives. This

17 Milton J. Esman, 'The Politics of Development Administration', in *Approaches to Development: Politics, Administration and Change*, ed. John D. Montgomery and William J. Siffin (New York: McGraw-Hill, 1966), 85–104.

dimension addresses the capacity-building of institutions to enable them to manage change, foster economic growth and support social development.

2. Operational efficiency: The second cluster deals with the efficiency of administrative systems in executing development programs. It covers issues related to bureaucratic performance, policy implementation, resource management and coordination between different levels of government and agencies. Operational efficiency ensures development plans translate into tangible results and reach the intended beneficiaries.
3. Socioeconomic and political context: This dimension examines the interaction between administration and a country's socioeconomic and political environment. It acknowledges that factors like political stability, social structures, cultural norms and economic conditions heavily influence the success of development administration. This cluster addresses how administrative systems must adapt and respond to developing nations' unique challenges.

These dimensions, identified by Esman, provided a framework to analyse the role of public administration in fostering development. They remain relevant for understanding the challenges of governance in developing countries.

The process of development planning

Developed and developing countries continually study and experiment with new techniques for national development planning. Effective national planning must incorporate several critical elements:

Local participation

Integrating national planning with regional small projects is a significant challenge. This can be addressed by applying systems thinking and systems theory to identify, measure and trace the inputs, intermediate transactions and outputs of a nation-state conceived analytically as a development-action system. A social accounting system helps structure the development planner's universe. Social variables must be part of the plan, involving researchers and practitioners from various public affairs professions, including engineering, education, agriculture and public administration.

Importance of public administration

Public administrators provide valuable insights into national planning with their extensive experience in research and teaching. They combine public affairs and social action. However, to be socially valuable, they must extend their focus to include applied economics and the sociology of change.

The performance of administrative systems in developing countries

Developing nations rely heavily on bureaucracies to deliver public services, which have expanded significantly to support social and economic development. However, failures in executing development programs and projects are often attributed to inadequacies, dysfunctions and frequent breakdowns in performance.

Problems

Common problems faced by administrative systems in developing countries include:

- Lack of coordination
- Deficient delegation
- Delays, overload, rigidity and authoritarian style
- Lack of initiative and acceptance of responsibility
- Condescending and abusive attitudes of bureaucracy towards the public
- Legalism and procedural formalism
- Institutionalization of corruption
- Overstaffing to counter unemployment

Administrative systems do not yield profitable returns economically or politically. Public Sector Units (PSUs) consume scarce resources and often run into significant losses.

Solutions to administrative weaknesses

More developmental research is needed to find solutions to improve the administrative capacities of developing nations. Limiting bureaucratic functions is a crucial suggestion from the Comptroller and Auditor General (CAG). Bureaucracy cannot manage everything. Therefore, delegating some activities from its control is essential. Solutions involve private enterprises, voluntary agencies and the market mechanism. Specific actions include:

- Eliminating price regulations
- Selling PSUs to private stakeholders
- Decontrolling foreign exchange
- Decentralizing decisions and responsibilities to lower administrative levels
- Divesting to autonomous agencies or local authorities
- Inviting capable individuals to associate with projects externally
- Strengthening interest groups to counterbalance the bureaucracy
- Enhancing bureaucratic capacity

- Training administrators in their social roles
- Increasing technical knowledge of public affairs
- Conducting training courses in managerial control techniques, performance budgeting, central procurement and position classification
- Offering courses in ethics and values

In summary, bureaucracy must strengthen its values rather than its weaknesses. Administrative machinery is crucial for implementing development plans in developing nations. The most important considerations are the intensive examination and analysis of institution-building as a process and the problem of choosing among alternative investments in institution-building efforts for development administration.

Approaches to development administration

Various researchers explored approaches dealing with successful development administration.

Systems approach

Saul M. Katz presented the application of the systems approach to development administration by undertaking development action as a process.[18] He began by giving the problems of devising an analytic framework for dealing normatively with development action.

1. Purposive: Development action is purposive and concerned with achieving consciously chosen objectives by following normative guides for attaining 'what ought to be'.
2. Complex interdependencies: Societal change comprises economic, political and social elements in a complex interrelationship. Development action must give methods to deal with the complex interdependencies of societal change.
3. Span: Development action takes place over time. Planned development purposes and goals are designed for specific periods.

The systems approach provides an organising principle for integrated interdisciplinary study to deal with the problems above in development action. This

18 Saul M. Katz, 'Exploring a Systems Approach to Development Administration', in *Frontiers of Development Administration*, ed. by Fred W. Riggs (Durham, NC: Duke University Press, 1970), 123–140.

organising principle can be summarised in terms of two concepts: perspective and application. Perspective means providing a coherent framework for describing general relationships of the empirical world as a unitary whole. Application refers to models developed to represent each level in mathematical terms, which can be applied to the development action.

Characteristics of the system of development action

The characteristics of the system of development action are discussed in the following (Figure 5).

1. Purposive: The system is purposive, comprising variables or components in some defined relationships to output and is differentiated from the elements in the environment. For example, the government can allocate funds via budgeting to various ministries to achieve specified targets.
2. Components: These can be acts, needs or targets instead of always being organisations or agencies. Following the same principle as in the first point, the financial flow between the government and its ministries comprises components in measurable quantities.
3. Interrelationship among the components: The components are in a dynamic relationship with inputs and outputs, maintaining a state of equilibrium, which can be static or dynamic. Static equilibrium is composed of constant values of the components, and it exists in 'closed' systems where the environment has no impact on the system. An 'open' system reaches a state of equilibrium from a state of disequilibrium when impacted by changes and becomes a dynamic equilibrium.
4. Homeostasis: When specific parameters in an open system are altered, the variables may experience a temporary disruption in how they interact. However, the system may self-correct after some time, and the variables could return to their original values. This is called homeostasis for selected variables, like the market mechanism of prices.

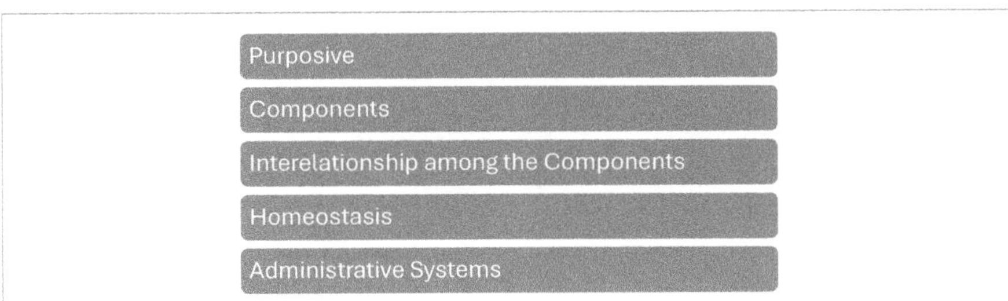

Figure 5 Characteristics of the system of development action

5. Administered systems: Four administrative functions must be carried out to achieve its purpose.
   - Decision-making: It involves selecting what is to be done from the available alternatives and how to do it.
   - Specification: It comprises detailed programming of decisions, designation of responsibilities and delegation of authority.
   - Communication: There must be a mutual understanding of goals, methods, problems and appropriate behaviour between the decision-maker and the implementing agency for the development action.
   - Control: Control refers to the progress of performance of the development programmes according to the standards set at the time of decision with permissible variations.

These four administrative functions are also called sub-systems.

Application of the systems approach to development administration

A development action system comprises four significant dimensions (Figure 6).

1. Outputs: These can be of two types. The first is the specifically defined development targets, and the second is routine tasks, like tax collection, irrelevant to development objectives.
2. Inputs: These are the technical developmental needs required to transform them into outputs. Inputs and outputs in a system, with its components and constraints, work towards a successful development plan and eventually

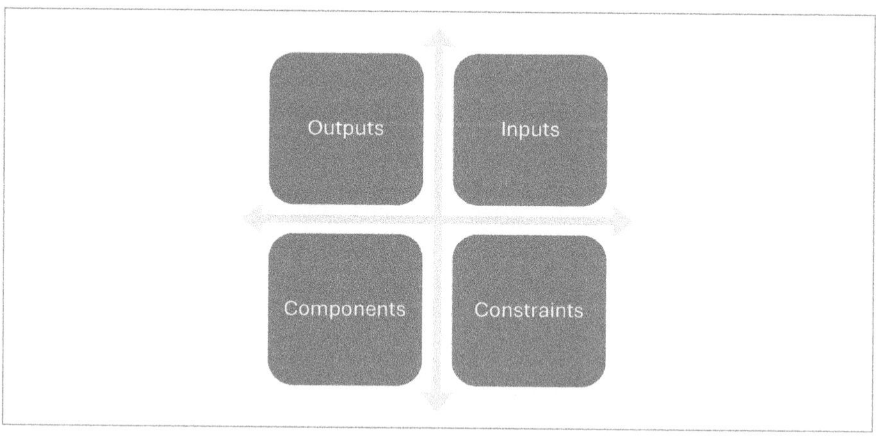

Figure 6 Application of the systems approach to development administration

towards development administration. The outputs in a system are the inputs that meet the technical development needs.

Generally, six inputs are required:

- Skilled manpower
- Finances
- Logistics (facilities for the physical flow of goods and services)
- Information (facilities for the physical transmission of data)
- Participation (of individuals and groups)
- Legitimate power (to enforce decisions)

3. Components: These are the characteristics of the entities that are related to each other in a system, such as the movement of trained manpower. The entities can be individuals or organised or unorganised groups. They can have further sub-systems.
4. Constraints: In a system, constraints define what is feasible. They can be of two types: technical and social. Technical constraints are based on technical procedures and relationships, whereas social constraints are the consequence of the social environment.

The preceding discussion described the normative pattern of a possible development action plan. The systems approach lays out a development action plan as a framework applied to development administration.

Governments as agents of developmental change

Governments are the primary agents driving developing countries' economic, social and political developmental changes. The second crucial component is the input needed to create facilities related to social and physical infrastructure. Development administration functions similarly to traditional administration in implementing policies, rules and norms but differs in scope, objectives and complexity. Development administration is innovative as it aims to bring societal changes to achieve developmental goals. Government activities have expanded from maintaining law and order to delivering a wide range of essential services, significantly widening the scope of development administration and requiring extensive coordination.

Systems approach and institutions

Government organisations perform developmental functions by ensuring that outputs are achieved by facilitating inputs through division of labour, coordination and stability over time. Organisations are technical instruments that act as means to achieve specific goals and hold value for individuals and their social environment. In development administration, these organisations must align with development objectives. Thus, development-related institutions must be established, leading to the concept of 'institution-building' by Harlan Cleveland.[19] These institutions should be viewed as systems with innovation, encompassing two main aspects: technical and institutional.

Advantages of the systems approach for development administration

The framework for development administration includes:

- Technical action systems
- The relationship of these systems to their social environment
- Four administrative functions

This framework forms a normative study of development administration. The utility of this approach is summarised as follows:

- Addresses many interrelated variables to understand and manage the complex interdependencies of societal change as an interrelated system
- System elements work towards achieving developmental action
- Selected system elements act as objectives to make them an empirical reality
- Systems can be put into dynamic terms to effect societal change
- Clearly defined systems remove obstacles to effective operation
- Quantification of variables and their relationships becomes a powerful tool for mathematical treatment[20]

Problems in the systems approach

The systems approach emphasises the general administrative mobilization and the decision to reorganise traditional government organisations or create new agencies. According to Katz, a systems framework provides administrative flexibility, highlights problems, suggests relative priorities and offers dynamic guidance. However, there are many problems in achieving development action, as shown in Figure 7.

1. Boundary: The systems approach defines the sub-systems and segregates them from their environment, but the boundary presents two difficulties.

---

19 Cleveland Harlan et al., *The Overseas Americans* (New York: McGraw-Hill, 1960).

20 Saul M. Katz, 'Exploring a Systems Approach to Development Administration', in *Frontiers of Development Administration*, ed. by Fred W. Riggs (Durham, NC: Duke University Press, 1970), 123–140.

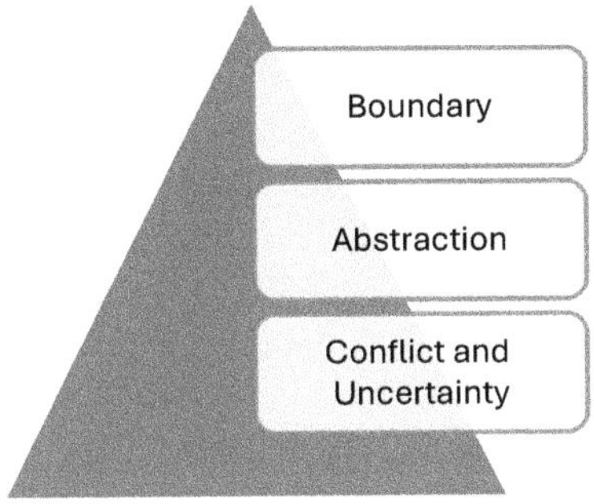

Figure 7 Problems in systems approach

First, there is a problem in identifying the criteria to make the boundary. Second, defining the conditions that distinguish a system is influenced by its goals or proposes due to the prevailing objectives not being achieved and there is no hope for the future. Additionally, the objectives have been defined by aping the Western developed countries unsuitable to the developing nations.

2. Abstraction: In the systems approach, the key variables must be identified to simplify a complex problem into measurable terms. Sometimes, it becomes challenging to break down an abstraction into measurable sub-systems.
3. Conflict and uncertainty: Development administration, comprising development action systems, is full of conflicts and uncertainties. The objective of change involves conflicts, and the outcome is majorly affected by how these conflicts are resolved. Another factor that is relevant in attaining the required outcomes is information. If the available data is inadequate and inaccurate, the result will also be adverse, thus becoming a problem in the systems approach application.

Hence, data collection methods must be developed to obtain precise and empirical information that will help identify and define the developmental objectives and make systems and sub-systems in relation to other variables. The systems approach can be an improved methodology to achieve and sustain developmental action plans.

Economic development approach

In no other field, orthodoxy lasts lesser than development economics. Today's general strategy for economic growth becomes tomorrow's barrier. A factor considered crucial in one schema becomes a highly dependent variable in another.[21] Economic growth and development are instrumental in development administration only to a certain point. Beyond that, other factors come into play, such as psychological, sociological and administrative factors. The economic strategy of development administration states that development administration is the product of a cross-disciplinary social science perspective with a substantial contribution of political scientists.

Economic development strategies: The 'big push' balance growth strategy

Ragnar Nurkse and Paul Rosenstein-Rodan proposed the 'big push' balanced growth strategy for developing underdeveloped countries. It stated that the low per capita income that prevails in developing countries is a consequence of the low levels of productivity that require increased capital formation. The demand for manufactured goods is small and the supply is used for subsistence. The cost of production is high, which decreases the levels of capital formation and hence, productivity does not increase, resulting in the status quo of poverty.

Reasons for low per capita income in developing countries

In a sequential form, the reasons for low per capita income in developing countries can be explained in a flow chart, as shown in Figure 8.

The strategy

The strategy for 'the big push approach' is as follows:

- A whole set of complementary units should be made, for example, by giving a stake to the workers in the factory or letting them buy the produce to solve the problem of demand and capital formation
- There can be a combination of investments which will prove profitable with substantial changes in production patterns, expecting no help from the market mechanism
- To achieve balanced growth, every component must be in quantity at the right time and place

21 The internal representation of the world is an organisation of concepts and actions that can be revised by new information about the world. Warren F. Ilchman and Ravindra C. Bhargava, 'Balanced Thought and Economic Growth', in *Frontiers of Development*, ed. by Albert D. Barnett and Fred W. Riggs (Durham, NC: Duke University Press, 1970), 75–90.

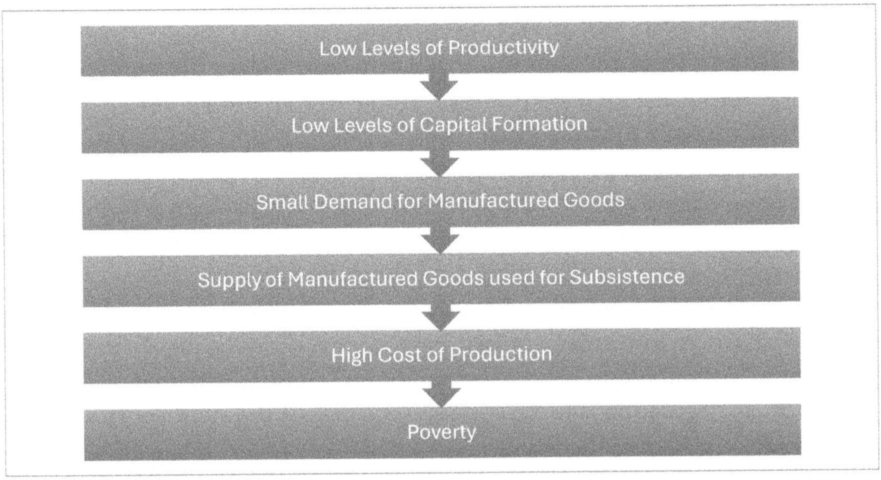

Figure 8 Reasons for low capita income in developing countries

Sources of capital for the 'big push'

There are two sources of capital for the 'big push'.

1. Domestic savings: The chief sources of domestic savings are taxation, higher profits and inflation. In a developing country, these means cannot be adequate, and it will have to rely on foreign capital inflows.
2. Foreign savings: Foreign capital inflows must be relied upon through government grants, loans and direct private investment. The 'big push' approach uses 'non-economic factors' in a developing country to improve various parameters. A political ruling elite that is strong, has a purpose and is united is required, along with a capable administrative system, to implement the big push strategy. It can be of two types.
   - Modernizing elite: The intellectuals, the military and the civil service form the first group that acts as an agent for balanced growth. There is government ownership and excessive control of the private sector.
   - Business sector: A small, closely-knit business sector can be developed to organise a whole complex of industries in a balanced growth pattern. The government must support the effort by giving incentives, providing easy business conditions, levying low taxes on the industries and the elite, investing in infrastructures, providing technical help and controlling the trade union.

The ruling elite must be narrow in membership to be effective; otherwise, the efforts will be diffused by the demands of many members. This implies that 'the few' are authoritarian while undertaking the 'big push'.

Problems in the 'big push' strategy

The big push strategy presupposes that a few ruling elites are planning and executing industrialization quickly. The problems that come in the way of the big push strategy are the following:

1. Effects of rapid industrialization: The big push strategy will push the economy into a phase of rapid industrialization in a society which is economically primitive. The implications are the following:
   - A large portion of the rural population will move towards the urban areas where industries are being set up, overloading the metropolitan areas with problems of housing, sanitation, medical attention, and so on.
   - Labour legislation will be complex because of the tight control of the trade unions over the labour force.
   - Rural people coming to the urban areas find adjusting problematic and general dissatisfaction spreads among the masses.
2. Regional disparity: The big push strategy will lead to regional disparity and must be one of the points in the strategy itself to be redressed. The adverse situation of regional disparity arises because of the prejudice of the ruling elite towards certain areas to become economic hubs.
3. Group demands: The new nations are averse to foreign capital through aid, loans or capital investment. In such a situation, higher rates of taxation and inflation put an unnecessary burden on the non-profit earning classes. The middle class gets frustrated, while the opposition can take advantage of the social situation arising from the big push strategy in the developing countries.

The theory of unbalanced growth by Albert Hirschman

Conceptual overview

Albert Hirschman proposed the theory of unbalanced growth in response to the big push theory, arguing that development is gradual. He suggested creating situations where people are forced to make decisions by deliberately unbalancing different sectors of the economy. When some parts of the economy grow, shortages in complementary parts create pressure to increase investments. Governments should build infrastructure to allow entrepreneurs to make profitable investments. However, infrastructure investments are often underutilised for long periods. Governments can encourage entrepreneurs to invest in productive activities, efficiently accelerating development using available resources by

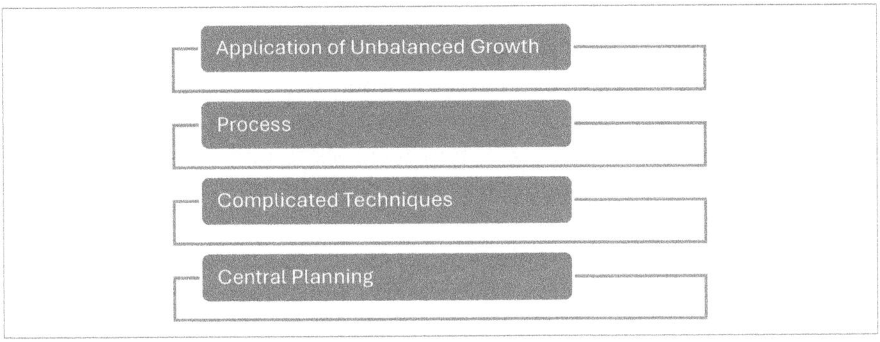

Figure 9 Elements of the theory of unbalanced growth

providing minimum infrastructure and offering incentives like tariff protection and subsidies.

Elements of 'the theory of unbalanced growth'

The elements in the theory of unbalanced growth are given in Figure 9.

1. Application of unbalanced growth: The theory can be applied to other issues, such as where to invest, which technology to use and how to keep the capital flowing.
2. Process: The initial activity selected must maximise the backward and forward linkages to maintain productive investments.
3. Complicated techniques: Complicated techniques must be employed to pay more attention to the maintenance of the factories, plant machinery, and the like.
4. Central planning: Central planning must be avoided because it internalises the external economies and diseconomies. Rapid growth can be achieved by aggressive industrialization.

Critical assessment

Even though the theory of unbalanced growth considers the issues plaguing underdeveloped societies, it has neglected the political, social and administrative factors. The problems of the masses are ignored because the dominant business class in the nation would monopolise power without government responsibility, leading to low wages, worker exploitation and inadequate investment in social infrastructure.

The bureaucratic approach

Ferrel Heady researched and proposed a bureaucratic approach to development administration.[22] He pointed out that the bureaucracies have been neglected and that no work has been done to describe and prescribe their objectives. Heady evaluated the comparative study of bureaucracies to understand the suitable role of bureaucracy in developing systems.

Characteristics of bureaucracies

Given below are the premises of bureaucracy which have been agreed upon universally:

Definition: A public bureaucracy is a form of large-scale organisation in nearly all world nation-states.

- Variations: The bureaucracies have common structural characteristics but vary depending on the adaptation to the political environment in which they exist.
- Western bureaucracies: The bureaucracies in developing countries differ from the classical model of bureaucracy found in developed Western countries.
- Lack of knowledge: The internal operating characteristics of bureaucracies in developing countries have not been studied much.
- The political role of bureaucracies: Bureaucracies are more prevalent in developing countries than in Western countries and affect political development.
- Guidelines for bureaucracy: These are needed in developing countries as they are highly involved in political processes.

Approaches used within the bureaucratic approach

Bureaucracies perform different functions in different political systems and must be studied empirically for a better understanding (Figure 10).

There is a combination of input and output functions. Input functions include political socialization, interest articulation and interest aggregation, while output functions involve political decision-making and execution. According to Riggs, developing societies show characteristics of a prismatic society and the 'sala' (see the Riggsian Model in Chapter 2).

Ferrel Heady reviewed what other thinkers said about developing countries' bureaucratic strategies. Some of them are as follows:

---

22 Ferrel Heady, *Frontiers of Development Administration* (Durham, NC: Duke University Press, 1970).

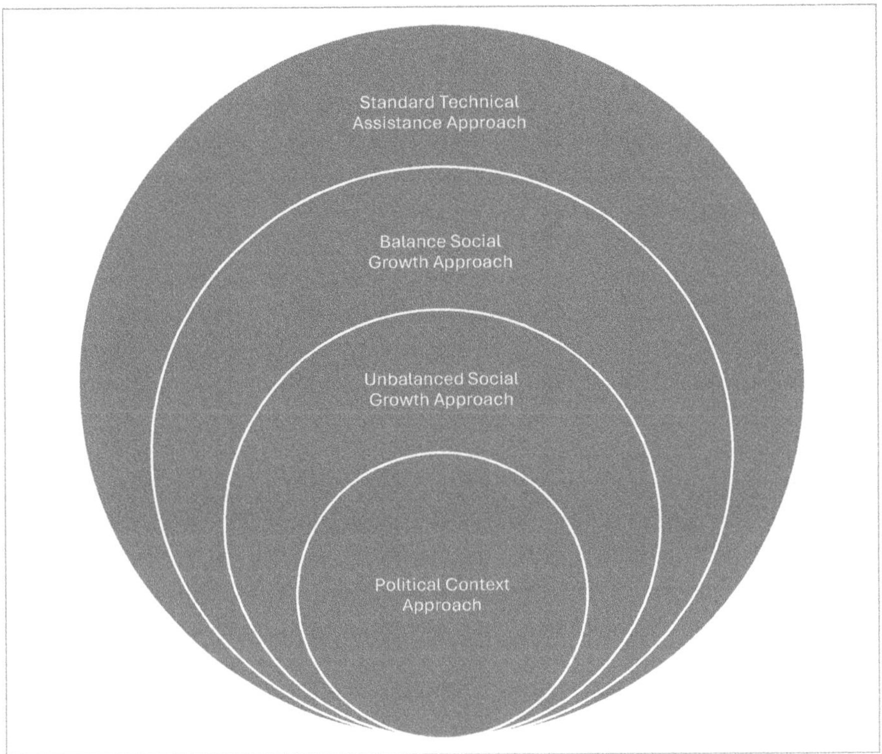

Figure 10 Approaches within the bureaucratic approach

1. Standard technical assistance approach: This approach proposes that administrative reform is intrinsically good without regard for its political consequences.
   - Assumption: Administrative improvement is desirable and must be done wherever feasible at the fastest rate possible. The bureaucracy must be as efficient as possible to operate in any political regime. Ilchman called it the 'administrative systems approach which is dominant in development administration in particular'.[23] The belief is that the administrative systems of the West can be adapted to developing countries selectively and with some modifications. Developing countries benefit by not facing the problems seen in developed countries.

23 Warren F. Ilchman, 'Rising Expectations and the Revolution in Development Administration', *Public Administration Review*, 25.4 (1965), 314–328.

- Growth in administration: Balanced or unbalanced growth in administration can occur depending on whether they urge simultaneous, phased, across-the-board reform efforts to upgrade administration broadly or prefer to concentrate on a limited number of critical segments of administration, expecting these to bridge the way for further advancement.[24]
- Outcome: Administrative development can be a prerequisite to political development, leading to rapid political and economic growth, but there is no specific prescribed method for different developing countries.

2. Balanced social growth approach: This approach emphasises developing bureaucracies from a historical and societal perspective. The bureaucracy must suit the development stage of the society. As a power centre, the bureaucracy must be appropriately related to other power centres in society. The fear is that bureaucracy might become more dominant and powerful than other political structures. Developing countries have powerful bureaucracies compared to other political institutions, making it challenging to keep them in check. Bureaucracy monopolises political direction, favouring group interests. If unchecked, the imbalance between the bureaucracy and polity increases, becoming counterproductive. The government must stop all aid to curtail the bureaucracy's power.
3. Unbalanced social growth approach: Scholars like Ralph Braibanti, Milton J. Esman and Ilchman proposed this approach. Development is highly differential, unbalanced or asymmetrical, with bureaucracy ahead of its competitors. Administrative reform from the West must be autonomous, regardless of political progress. Given below are the major components included in this approach.
   - Improvement Needs: There is an urgent need to improve administrative capability to prevent it from being counterproductive. Aid from developed countries can be withheld, or administrative capability can deteriorate.
   - Regular Feature: Unbalanced growth is regular and political development occurs differently.
   - Transitional systems: Transitional systems can be tutelary, primarily military, etc.
   - Civic Culture: Shallow civic culture has little say in the political process. The administration needs to be more responsive.
   - Improvement programme: A rapid, broad-based, long-term administrative improvement programme might tilt the balance between bureaucracy

24 Ferrel Heady, *Frontiers of Development Administration* (Durham, NC: Duke University Press, 1970).

and polity but would improve the social order and society. Instead of controlling bureaucracy, administrative reform must be sought and implemented.

4. Political context approach: Edward W. Weidner (see Chapter 2) and Joseph LaPalombara proposed the political context approach, stressing the bureaucratic approach according to a specific situation. Reforms must be based on the political configuration of the society linked to the nation's political goals.
   - Imbalance: If bureaucracy is more developed than the polity, there will be an imbalance, but administrative reforms must continue until balance is achieved because development cannot be stopped.
   - Shift Realization: New nations are not ready for bureaucratic–polity balance due to complex problems in resettling new institutions and systems. The overall environment of the concerned developing countries must be studied and analysed before prescribing improvements for the bureaucratic structure. Western approaches cannot be applied verbatim and must be modified per the situation.

To sum up, in Wilson's words, 'administrative development cannot take place in a political vacuum'.[25]

The present status of the development administration

An exploration of the challenges of economic growth in middle-income countries can be found in *World Development Report 2024: Economic Growth in Middle-Income Countries*. Using the World Bank's income classification, the world currently comprises twenty-six low-income, 108 middle-income and eighty-three high-income economies. Middle-income countries, with about 75 per cent of the world's population, now account for about 40 per cent of global economic activity, fifty percent of the world's extremely poor people and 60 per cent of global CO_2 emissions.[26]

Implementing developmental policies and projects is now termed 'governance', and doing it effectively and efficiently is called 'good governance'. It covers indicators such as voice and accountability, political stability and absence of violence, government effectiveness, regulatory quality, rule of law and control of corruption.

25 Fred W. Riggs, 'Relearning an Old Lesson: The Political Context of Development Administration', *Public Administration Review*, 25.1 (1965), 70–79.

26 World Bank, *World Development Report* (2024), <https://www.worldbank.org/en/publication/wdr2024> [accessed 2 July 2024].

Over nearly seven decades, development administration has achieved slow-paced success. Its nature has evolved into 'better governance' combined with sustainable development. World indices of human and economic development show increased per capita income and a rise in the Inclusive Development Index, though the situation remains dire. The World Bank defined sustainable development goals (SDGs) and inclusive development goals to end poverty and share prosperity. Renewed and innovative efforts are required to realise these goals of good governance and corruption in developing countries must be checked by strict implementation of anti-corruption laws.

In the past two decades, remarkable progress has been made in global socio-economic indicators, with one billion people rising above the poverty line. However, globalisation has harmed the equitable distribution of benefits, resulting in capital inflows to developing countries, boosting their economies and opening their markets to the West, thereby increasing inequalities. Countries in Africa, Asia and South America face destabilization and violence, hampering development and poverty reduction goals. Although developmental programmes initiated by various international agencies have been appropriately framed, the benefits have not percolated to the poorest. Corruption in developing countries has hindered development administration from achieving its goals.

Conclusion

Development administration has come a long way after the Second World War. The subject has been researched extensively, and its profile has changed over the last seven decades to what it is today. Various scholars have established the fact that developing countries need their models to be effective and must be self-sufficient so that they do not have to depend on Western aid; instead, they must be independent in their political, economic and social development and enter the international market to trade on their own.

In conclusion, the study of development administration has offered valuable insights and substantial contributions to the field of public administration over the decades. It provided a robust framework for understanding the complexities of governance, planning, and implementation in varying socio-economic contexts. The discipline's exploration of bureaucratic structures, political dynamics and developmental strategies has enriched our comprehension of how public administration can drive progress and improve governance. However, as the field evolves and new challenges and paradigms emerge, it is evident that the era of development administration as a distinct and separate domain has drawn to a close. While its relevance and influence were significant while it lasted, contemporary global administration and governance shifts call for a broader, more integrated approach that transcends the traditional boundaries of development administration and moves into sustainable development governance. This transition reflects the dynamic nature of public administration, where adaptability and innovation remain essential to addressing the ever-changing landscape of global governance.

CHAPTER 2

Theories of Development Administration

An International Perspective

The term 'development administration' was coined by a civil servant of the Indian Civil Services, U.L. Goswami, in his 1955 article 'The Structure of Development Administration in India'.[1] The subject gained importance when scholars such as George Gant, F. W. Riggs, Edward Weidner, Ferrel Heady and many others researched the concept of development administration. The subject developed its identity after the leaders realised in the post-Second World War period that the newly formed and freed underdeveloped countries needed development planning and strategy to move towards the development path. This chapter will explain the significant theories in development administration that established the models and how they dwindled by the 1980s. As the emphasis on sustainable development and its governance increased, development administration's relevance became redundant. However, it is vital to understand how the development administration models that failed in the 1980s were made. They had to give way to sustainable development when the need arose to take environmental concerns seriously.

Karl Marx and administration

Karl Marx (1818–1883) was a German philosopher and a socialist born in Prussia. He was expelled from Germany and France in 1844 because of his socialist and anti-capitalist views and writings. He was exiled to London and stayed there for the rest of his life. He wrote several books and articles with radical views and theories linking all the socialists from Europe. He established 'The Communist League' in London and gave the most iconic of slogans, 'Workers of all lands unite!'[1]

Economic historians universally accept that development administration emerged in the post-war period, but understanding Karl Marx's views is relevant here. His theory did not discuss development administration directly, but the expected result was the betterment of the common population. His views and theories were against bureaucracy and capitalism, but his goal was

1 Karl Marx, *Karl Marx on Administration* (Warangal: Modern Printing Press, Karkatiya School of Public Administration, 1985).

the development of society, which is one of the objectives of development administration. He remarked about development administration in several books. Marx understood that bureaucracy and public administration were essential elements for a government's function, but considered bureaucracy a dangerous institution. Marx wanted it removed from Russia after the overthrow of capitalism but could not do so. Marx may have criticised bureaucracy to the hilt, but in the end, it is indeed required to organise, facilitate and maintain the flow of governmental activities for development. The extent of his criticism of bureaucracy can be seen in his own words:

> The bureaucracy has the state, the spiritual essence of society, in its possession as its private property. The general spirit of the bureaucracy is secret, the mystery, preserved within itself by the hierarchy and against the outside world by being a closed corporation. Hence, authority is the basis of its knowledge, and the definition of authority is its conviction. For the bureaucrat, the world is a mere object to be manipulated by him.[2]

Marx believed that inequitable economies exist due to capitalism. The discussion here would be limited to Marx's contribution concerning administration as it is vital to understand his views on the impact of development administration. Despite Marx's antagonism towards bureaucracy, it is a bridge to cover the gap between the rulers and the subjects. This study is based on one of his 'early writings' supplemented by Shlomo Avineri as a handbill to be distributed among the students to strengthen the Marxist framework of administration.[3] Figure 11 shows the basic premises of Marxian views on administration and bureaucracy.

The gap between the rule of law and its execution

There is a wide gap between a monarch's decisions, laws, rules and regulations and their execution. The leading cause of this gap is the bureaucracy, which subsumes the powers of the judiciary and the police. Marxian theory emphasises the role of class struggle and the dominance of the ruling class over the working class. In the context of development administration, this can be seen in how development policies often reflect the interests of the elite or ruling class rather than those of the marginalised or poorer sections of society. Development administration might prioritise projects that benefit the wealthy or those in power, perpetuating existing inequalities.

---

2 Friedrich J. Carl, *Responsible Bureaucracy: A Study of the Swiss Civil Service* (Cambridge, MA: Harvard University Press, 1932).

3 V. A. Panandikar Pai and S. S. Kshrisagr, 'Bureaucracy in India: An Empirical Study', in *Administrative Theory* (New Delhi: Indian Institute of Public Administration, 1984).

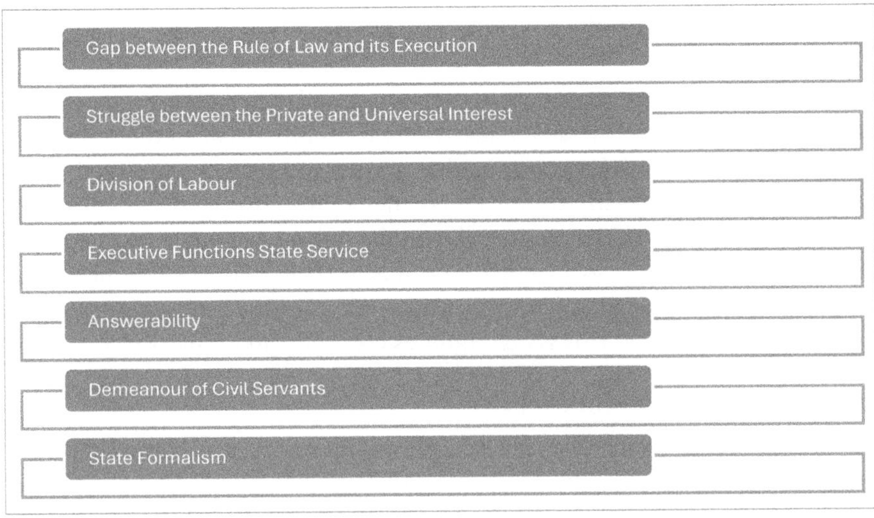

Figure 11 Premises of Marxian views on administration and bureaucracy

Interests

The universal interest of the people is looked after by the civil society (CS), while the state administers these interests in its power, for example, in municipalities, corporations and private property management. The state handles the recruitment of these organisations. The universal rights are maintained by the officials with executive power, such as the executive civil servants and the higher advisory officials who are members of various committees. They are in alliance with the supreme heads and are in direct contact with the monarch.

Struggle between the private and universal interests

There is a struggle between the private and universal interests, pitted against the organisation or the state. The corporations also acquire a state-like attitude and mindset to meet their ends. The citizens consider the state their substantial authority because it cares for their needs and welfare.

Division of labour

Division of labour also occurs in the executive business. The bureaucrats aim at:

- Governing the civil life from below as it is concrete at the bottom
- Dividing the government's administrative system into different branches manned by notable officials
- Convergence of all the departments from the above leads to convergence of general supervision in the supreme executive

Nature of the executive functions and recruitment of the executives

Executive functions are carried out by individuals and their authority comes from the decisions they have already made. There is no link between the individual and the executive functions, and neither are the individuals appointed based on merit in knowledge and ability. It becomes a subjective act if the candidates are nominated only for the service. The public officials act due to the objective authority of the crown's sovereignty. This means there is no need for nerds and geniuses in the bureaucracy. The bureaucracy is a collective group with the same level of intelligence. So, this system ignores creativity and innovation.

Requirement of the state service

The state service requires that the officials sacrifice their selfish and subjective ends and find satisfaction in discharging their public duties. This concept assumes that once there is assurance of satisfaction of needs of the people, the external needs would not be able to interfere in performing public functions.

Answerability

The state derives security against the misuse of power by public officials and the ministers by the authority given to societies and corporations. The officials of the state are answerable to their department heads.

The demeanour of civil servants

They become dispassionate and polite due to their education and training. The size of the state also influences the officials' nature to adopt universal interests. They form the middle class, which has the consciousness of righteousness. The middle class is politically conscious and well-educated, but this can only be achieved if the state has organic unity. A state with organic unity is one in which authority is sanctioned to various agencies with particular interests relatively independent of the army without personal arbitrariness.

Administration bureaucracy

Hegel, Marx's companion in his writings and views, stated that since the police and judiciary are a part of civil society, the executive, consequentially termed bureaucracy, is the administration.

Bureaucracy as the state formalism in civil society

Hegel wrote more emphatically than Marx on the concept that the corporations, the representatives of the people, form the state formalism of the civil service. Bureaucracy in the corporation must abstain from individual interests to maintain the corporation mindset. Marx stated, 'The corporation is civil society's attempt to become the state, but the bureaucracy is the state'. In development administration, bureaucratic structures might be used to maintain control over resources and decision-making processes. This can lead to development initiatives that reinforce existing power structures rather than challenging them or fostering genuine grassroots development.

Control

Civil society is protected by the hierarchical organisation of the bureaucracy. The conflict between the bureaucracy and the corporation remains unresolved while the possibility of struggle would act against the bureaucracy being authoritarian. Marx's concept of alienation, where workers become estranged from their labour, can be related to how individuals in developing countries might feel disconnected from development projects. Bureaucratic procedures can be impersonal and detached, alienating those supposed to benefit from development. This can result in a lack of engagement and participation from local communities.

Bureaucracy: The imaginary universality

Marx held similar views to Max Weber's about the bureaucratic 'ideal type' focusing on dividing functions and hierarchy. He recommended the abolition of the state by destroying the bureaucratic apparatus. The bureaucratic dimension of political reality offers a criterion for assessing different political structures. He said that human beings have become mere objects of manipulation. For instance, in France, bureaucracy had become an instrument of the ruling class, however much it strove for its power. Marx argued that freedom existed in converting the state from an organ superimposed upon society into one completely subordinate to it. Marx remarked on independent bureaucracy within capitalist corporations. The state must reject its origins in the material world.

To sum up, Karl Marx wrote a lot about its role and the nature of the state. He highlighted the ills of Western capitalism, namely the exploitation of the workers and people experiencing poverty. He later presented the conflict theory, which claimed that society is always in a continuous conflict because resources are limited and there will always be competition. His theory states that social order can be obtained only by force and domination instead of conformity and consensus. We can critically examine and address the power imbalances and inequities inherent in many development policies and practices, advocating for a more just and inclusive approach to development by applying Marxian bureaucratic ideology to development administration. However, history shows that his theories failed when the USSR disintegrated in the 1990s, even though it always kept the common masses in mind for any state activity.

Marx and India

In 1853, a Bill regarding the Charter of the East India Company was introduced in the British Parliament. Karl Marx wrote three articles concerning the situation in India: 'British Rule in India', 'The East India Company: Its History and Consequences' and 'The Future of Consequences of the British Rule in India'.

He compared the socioeconomic self-sufficiency of the village system in India before the British came to India with the status of villages after the British administration had taken over. He cited examples of the export of gold, silver, muslin, and silk to England from India, which increased to very high levels in 1837, but India was importing cotton yarn to be spun into cloth.

The British administration was exploiting the workers, but the Indians were unable to fight back. Plunder and loot kept on increasing. Marx remarked that India was ridden with many complex problems, such as the caste and religious divide. He deeply loved humanity and wrote extensively to explain the issues and probably work out solutions. He was overtly sympathetic towards the workers who had to do rigorous labour for necessities and got paid significantly less. He held atheistic and anti-materialistic views. In 1942, J. R. D. Tata said, 'A capitalist has to evaluate how far it is possible to accommodate socialist demands without compromising on the basic features of Capitalism'.

Critical assessment

Marx was criticised for overlooking the bureaucracy's essentiality in modern political and socioeconomic life. He always suggested overthrowing the bureaucracy by the proletariat. He did not write specifically about development administration. Still, socialist ideals can be included in capitalist principles to the extent that they deal with the socioeconomic problems of developing countries. One such initiative India and many other nations are practising is corporate social responsibility (CSR), in which corporations must keep aside a percentage of their financial resources, fixed by the government, for social welfare.

Bureaucratic model in development administration

Bureaucracy is the foundation of a functional administrative system common to all modern nation-states. Any administrative work in a government requires personnel with legal authority within the bureaucracy. The origin of the word 'bureaucracy' must be traced to understand its meaning. Although the historical dating is not definite, it was first used by a French minister of commerce in the eighteenth century. The Germans modified the word to 'burokratie' in the nineteenth century, and Max Weber extensively used it in his writings. Thus, 'bureaucracy' is more of a concept than merely a group of individuals with authority.

Bureaucracy: Types of authority

Various scholars have defined bureaucracy within governments, emphasising its structure and behavioural traits. There are three essential structural features of bureaucracy:

- Hierarchy
- Differentiation or specialisation
- Qualification or competence

The features of bureaucracy cannot be discussed without considering Max Weber's concept. He divided authority into three types, as noted in Figure 12.

1. Charismatic authority: This is characterised by permanence and often unstable as charisma of the leader fades or followers are lost. It resembles kingship or chieftainship.
2. Traditional authority: This is based on religious, cultural, or military norms, emphasising routine work.
3. Legal rational authority: It is mostly predominant in modern states in which authority rests on laws and norms, characterised by four attributes:
   - Conduct: Bureaucrats must follow the rules and laws.
   - Jurisdiction: Departments handle specific issues hierarchically, with grievance appeals allowed.
   - Specialised training: Staff must be trained in their field of specialisation.
   - Means of production: Staff is unrelated to ownership of production or administration, with no rule changes allowed.

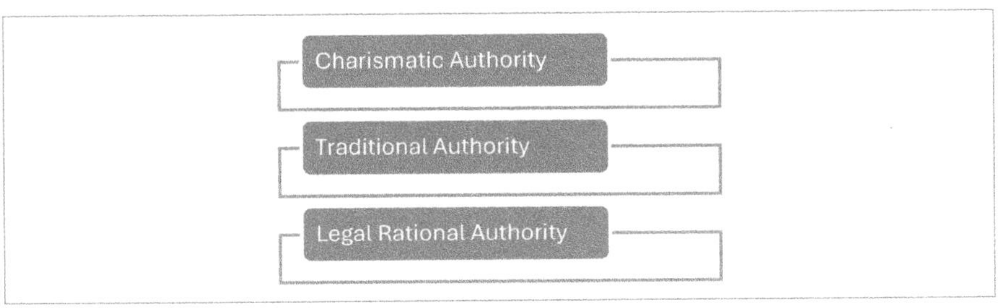

Figure 12 Types of authority

Weber's bureaucracy promotes legally free individuals and hierarchical staff hired by merit, governed by discipline and rules, and efficient in stable environments. However, he did not consider the unstable environments of developing countries.

Bureaucracy and development administration

A structured hierarchical administrative system may succeed in some nations but often fails in development contexts requiring teamwork. Theories from the 1970s and 1980s for implementing development programs failed, leading to the growth of development administration as a discipline although it miserably failed. Weber's emphasis on value-neutrality overlooks the importance of understanding 'who' or 'what' in rural development, where officials are often perceived as arrogant and indifferent. While high-level bureaucrats may be rational, the gap between them and field agents hinders development administration. Models by Weidner and Riggs also failed due to their vague macro approaches. They presented problems accurately but did not offer workable solutions for developing countries. Expectations and goals in developing countries have shifted towards industrialised and modernised development. Almond Gabriel highlights the role of bureaucracy in this context by stating 'the degree of bureaucrats' involvement in political activities varies inversely with the level of political development, differing from country to country'.[4] This implies that the relationship between bureaucracy and politics varies across societies.

Bureaucracy and politics

Development administration requires bureaucrats to be politically involved in effective processes, though this role is counter-rational. Bureaucrats must maintain good relationships with politicians while staying apolitical. This dynamic is exemplified by 'delegated legislation,' involving structural-procedural and policy initiation roles. Bureaucracy must be reformed to adjust to changing development administration profiles. If bureaucrats remain detached from reality, development efforts will suffer.

Behavioural model of development administration

Although foundational, Max Weber's model of bureaucracy is criticised for its impracticality in modern contexts. Critics argue that no bureaucracy can be entirely rational and neutral, as it is influenced by ecological factors like cultural, political, economic and social contexts. These external influences can lead to

4 G. A. Almond and S. Verba, *The Civic Culture: Political Attitudes and Democracy in Five Nations* (Princeton, NJ: Princeton University Press, 1963).

inefficiency and susceptibility to corruption. Weber's rigid structures may also stifle innovation and adaptability, which are crucial in today's dynamic environments. Consequently, modern public administration often seeks more flexible, responsive, and context-sensitive approaches. Hence, a behavioural approach emerged, which was proposed by various scholars such as Carl J. Friedrich, Alwin and Michael Crozier. They considered behavioural characteristics essential to achieve goals in development administration, although the behavioural Model was also criticised for being ridden with red-tapeism, procedural complexities and subjectivity.

The behavioural approach developed in the 1930s and 1940s (the neoclassical period) considered the behavioural aspect of bureaucracy among social environments. Thinkers such as Mary Parker Follett, Herbert A. Simon and Chester I. Barnard laid the groundwork for the behavioural approach. Herbert Simon argued that administrative behaviour involves not only organisational rules and procedures but also the individual and collective behaviour of people operating within administrative environments. Simon emphasised that decision-making in administration is influenced by both personal choices and group dynamics. He believed that understanding how individuals act and interact with each other, as well as how they are influenced by the administrative system, is crucial to analysing the behaviour within organisations.

Thus, Simon's theory integrates both individual actions, such a decision-making processes, and collective behaviours, including how groups function in administration, within the broader framework of administrative systems. This idea forms a key part of his work on administrative behaviour, where he explores how real-life human behaviours impact the effectiveness of organisational structures and processes.

The behavioural scientists never actually talked about their approach concerning development administration but condescendingly mentioned the developing countries here and there. Some major constituents that the behaviourists proposed about administration and bureaucracy are briefly discussed.

1. Public policy: The research in public policy transitioned from its classical stance to being behavioural in nature. Researchers began data collection on a larger scale to understand the administration issues with new comparative techniques. In the 1960s, there was a shift from irrelevant studies to their being relevant, called the post-behavioural approach. Policy relevance became the watchword for research to study the real issues, but it was primarily done for the Western world. They studied public bureaucracy from a behavioural perspective.

2. Emphasis on the individual: The behavioural approach focuses on the individual behaviour, which is influenced by one's inner self and the exterior environment. Understanding human resource behaviour makes one appreciate the thinking patterns influencing their external behaviour, motivation and morale. It can only be derived from this approach that the planners and policymakers of developing countries consider the various dimensions of human behaviour to understand the expectations of the masses. Developing countries studied the psychological theories of David McClelland and Everett Hagen and applied them to their systems. McClelland and Hagen were impressed with the concepts of individual personality traits and society's overall general psychological attitudes. However, they were criticised because of their condescending attitude towards third world countries, which characterises them as having low achievement needs, lack of motivation, low morale, and lack of self-confidence.
3. Behavioural approach as a social change agent: Researchers developed the concept of human behaviour and personality traits and suggested that their study can be used to understand a society's social system and structure. The observations can be used to make a model to initiate social change. Human behaviour can be altered by introducing education technology, where the Western metrics of 'education' remain contested, as well as modern industry and mass media.
4. Criticism of bureaucratic models: The behaviouralists considered the bureaucratic model rigid and incapable of implementation as it lacked the inclusion of human behavioural reactions and ideologies, which must be considered when formulating policies and planning for development. In India, bureaucracy is inefficient and ineffective in implementing development administration as it has grown in size and is too self-absorbed to understand the public sentiment. The bureaucratic model, however, cannot be abandoned. Bureaucracy must be reformed by adding the perspective of human sensitization for development.

Critique of the behavioural model

The behavioural model could not present a substantial practical methodology for implementation, especially in developing countries. It stated that socioeconomic, psychological and political factors affected all human behaviour. It did not include the issue of values or ethics. These factors must be empirically researched to find solutions to administrative problems. Nonetheless, they enriched public administration research by comparative research studies in behaviour. In the recent decade, there have been approaches such as the public choice approach and new public management, which consider the issues of developing countries and present a viewpoint by theorising. They also had many weaknesses, leading to the emergence of the good governance model, which has limitations as it can be applied only to small groups of people or associations and not to administrations or bureaucracies.

Weidner model

Edward Weidner (1921–2007) was the founder of the University of Wisconsin – Green Bay and served as the Chancellor of the university for twenty years after contributing eight books and over 100 papers. Although an Indian scholar and civil servant, U.L. Goswami, in the Indian Administrative Services, coined this term in his article 'The Structure of Development Administration in India' in 1955, Edward Weidner gave the concept of development administration. He began his premise by criticising the traditional administrative theory, which focused more on means than ends. He stated, 'Good administration and human relations have become ends in themselves, quite apart from the achievement of other values they may facilitate'. Weidner formulated different development models based on directional growth, system change, and planning. He advocated transformation administration to bring about social change. This can be achieved by reforming and revitalising institutional mechanisms to attain action-oriented development and administration.

Definition

Weidner defined development administration as an 'action-oriented growth, change in the systems and planning' focusing on societal changes. He considered development administration in the government as 'the process of guiding an organisation toward the achievement of progressive political, economic and social objectives that are authoritatively determined in one manner or other'. He rendered development administration as innovating for development, especially the effectiveness of a bureaucracy. It helps to steer the direction of nation-building and socioeconomic progress. Development administration moves towards an increase in the rate of development by determining the conditions in which development can take place.[5]

The eight models formulated by E. W. Weidner are discussed next.

5 *Development Administration in Asia*, ed. by Edward W. Weidner, (Durham, NC: Duke University Press, 1970).

1. The ideal of a planned directional growth with system change: Ideally, Weidner conceptualised the first model by formulating plans as a continuous process to achieve growth in the desired direction concerning socioeconomic progress. Most Southeast Asian nations began formulating 'Five-Year Plans' to do so but failed in their performance and achievement of targets because the outcome was not focused upon. The developing nations made innovative changes in their planning for growth, but they failed due to inefficient administrative and political systems.
2. The short-run payoff of a planned directional growth with no system change: Weidner progressed to the second model, 'short-run payoff: planned directional growth with no system change' after realising the limitations of the first Model. He stated that since political or administrative change is impossible for the system, the LDCs must focus on short-run payoff, maximising development in the existing system. He gave four scenarios in which this Model can be adopted. This Model can be applied to reap short-run payoffs in the following cases:
   - A leader prefers to maintain a balance between the forces instead of bringing about unconventional and innovative changes.
   - There is a lack of technical assistance or availability at a low cost without any system change.
   - Political destabilisation
   - There is a demand to show results and an increase in consumption

   The common factor in all the cases mentioned above is the presence of political will and the resolve to facilitate development with well-formulated plans and projects, which must be executed with cost- and time-effectiveness. Weidner stated in his book *Development Administration in Asia* (1970):

   > An example of this Model is the promotion of native arts and crafts, such as the cottage industry programmes in India and Pakistan. These programmes were adopted as a part of the plans. They did not disturb the economic or social structures and, in turn, realised economic gains for the country's concern and peasants or other participants. These gains have been small.[9]

   However, 'planned directional growth with no system change' does not essentially mean that there can be no variations in ramping up the coordination training of civil servants or adding agencies to bring about effectiveness. There can be considerable increases in economic growth, but the growth will become stagnant after a certain point.
3. Thelong-run payoff of planned system change with no directional growth: Weidner states that in this model, as its name suggests, there will be a planned system change with no directional growth. Growth is considered

the direction of modernity, nation building and socioeconomic progress, which could be for the short or long run. According to Weidner, system change does not essentially culminate in growth. Moreover, it could have a contrary result. System change is distinct in nature in different political and administrative systems. It becomes quite complicated to monitor a system change if the society has its particular strong individuals or groups with high regard for their system and are not open to change.

4. Failure of planning with no growth or system change: In the 1960s, various scholars and economists believed that the Westernised concept of development administration and planning had not been instrumental in increasing growth or system change, making it a failure in developing countries and LDCs. Weidner illustrated this point through the example of Vietnam, where budget agencies were reshuffled. An innovative director was appointed to head the central agency of budgeting, where he trained the officers on how he realised it was necessary for effectiveness. This was achieved within a year, leading to increased output. However, the director was removed from his position because the system changes within the agency necessitated major changes in the government, which the government was not ready to do.

   Weidner gave four relevant propositions based on the Vietnam experience.

   - Threat of innovation: Innovation must not be perceived as a threat to the security of employees or the personnel system to make it successful. Weidner gave the example of India, which has a robust personnel system of civil services, but very little has been accomplished.
   - Ripple effect: System changes in one agency orchestrated by an innovative head do not limit themselves to that agency. It also has a rippling effect on other agencies, and they must make changes. So, there is resistance to any system change.
   - External agencies: This proposition is included in the second proposition, and it can be derived that there is strong resistance to change from the employees of the external agencies if the system changes extend to them.
   - Government-wide change: Lastly, Weidner proposed that innovation, short of government-wide change, will be inhibited to the extent that central controls are present. Weidner concluded that system change was challenging and was a failure if a macro picture was visualised.

5. Environmental stimulus of an unplanned directional growth with system change: This Model specifies that 'by deliberate actions, we can bring differentiation, coordination and growth'. Weidner said that modernity, nation-building and socioeconomic progress concerning high differentiation and

coordination can happen in an unplanned way, too. The planning process requires the determination of the following:

- Goals
- Priorities
- Programme of action
- Implementation devices
- Control

Weidner gave some essential points of consideration.

- Secure political environment: Completing the planning process is extremely difficult because nationwide developmental change requires a secure political environment.
- Decentralised innovative leadership: Weidner differentiates between planned and unplanned change, pointing out that most of the changes take place due to the reactions of a leader or a group to the environmental factors in society. Systematic, planned change at a nationwide level from the centre might be challenging but innovating leaders usually bring it about in a bureau or an agency.
- Adaptive system changes via bureaucracy: System change can be carried out by the bureaucracy at its level practically in their agency or department without any obligation to planned development.
- Emulation: A lot can be achieved by emulation via simulation from other models of foreign countries or agencies, given that there is a desire for modernity, nation-building and socioeconomic progress. Hence, without any significant planning, environmental stimulus can bring system change via environmental factors, decentralisation, adaptation and emulation.

6. Pragmatism and unplanned directional growth with no system change: This model is quite like the fifth model, which is rooted in decentralised system change with emulation and adaptation. The sixth model emphasises some development as an essential supplement to an extensive programme of planned change without any system change. It would be a natural progression towards a social change in the society.
7. Crisis and unplanned systems change with no directional growth: The crisis model relates to an unplanned system change without any directional change occurring because of the following:
   - Decentralised initiatives
   - Competition
   - Adaptation
   - Emulation

   This model works in unplanned system change without any decentralised growth, for instance, in crises such as war and post-war periods, formation of new nations, emergencies, people taking refuge, famines and other disasters. In such cases, on-the-spot adjustments must be carried out for the four indicators of decentralisation, competition, adaptation and emulation. These adjustments can also lead to major system changes, and in the case of extended crises, the adjustments can also be unplanned.

8. Static society with no plans, no change: The eighth model given by Weidner relates to totally or partially static societies. Similarly, there can be pockets of change or resistance to change in developing nations. According to this model, crises can lead to adverse consequences, contrary to the seventh model. There can be confrontation acting as an obstacle to widespread change.

A review of Weidner's models

Weidner has given eight extensive models for almost every conceivable situation. Developing countries need socioeconomic progress to grow in all areas, but each country has different conditions and norms that need to be considered before applying a model; otherwise, there can be adverse consequences, as mentioned by Weidner. He laid out the conditions and strategies to facilitate development. To sum up, Weidner highlighted three points which need attention:

- The end results of all government policies and goals must be modernity, nation-building and socioeconomic progress.
- Various combinations of conditions in society must be considered to increase development output.
- There must be a continuous research effort to reach a development model for a particular country.

Problems in Weidner's models

Weidner's contribution to development administration can in no way be undermined. He researched extensively to formulate the eight models. The problems faced by governments and administrations while adopting them have been discussed further.

- Application: It is not at all easy to eventually arrive at one development model for a particular country as each country has different problems. Even out of these eight possible scenarios, a combination of a few might work for a country as it is difficult to predict if the developmental outcomes of a particular model would be positive.
- Goal identification: This is a strenuous and complicated task in any country because of complex developmental change situations.

- Operational framework: Weidner has neglected and missed out on providing an operational framework laying out the type of administrative change required to transform a non-developmental system into a developmental system.
- Western bias: Weidner's models, as other models given by Western scholars, suffer from Western bias stressing on emulation. The socioeconomic conditions of a specific country are the key to develop a model for development planning in a country but this factor was ignored totally.
- Development criteria: Weidner based his models of development on three basic goals of nation-building, socioeconomic progress and modernity. It believed that economic growth would be an answer to all problems, whereas it has been established today by various international agencies and scholars that economic growth can only be a part of development planning. There are specific indicators like the sustainable development goals to measure the achievements with respect to inclusive development in developing nations.
- No recognition of participatory approach: Although citizens must be asked about their expectations from the administration and the government, their engagement and participation was totally neglected. All work is left to the bureaucracy which is a serious concern as the bureaucracy in developing nations is ridden with authoritative nature, corruption and red-tapeism. Although Weidner laid stress on administrative reform, he left political reform alone. Since administrative works happen in a political environment, there can be no positive change ultimately if the political environment is corrupted, insecure and unstable. CS is a major stakeholder in development administration but was ignored by Weidner in his models.
- Emulative administrative structures: The developing countries have emulated administrative structures because they were mostly ruled by imperialist forces and had to follow their administrative structures without having any choice in it.

Nonetheless, Weidner's models provide a keen insight into achieving the goals of development administration as they started research work on finding development models specifically for developing nations.

Riggsian model

Fred W. Riggs (1917–2008) was a political scientist and a trailblazer in formulating administrative models and theories. He is mostly known for his Riggsian Model which he theorised after studying intergovernmental administrations and cross-cultural societies. In his ecological approach to public administration, F.W. Riggs studied how social, cultural, historical and environmental factors influence administrative processes, highlighting that Indian administration is dynamic

Figure 13 Agraria industria model

and transforming. Riggs also developed the agraria-industria model, classifying societies into agrarian and industrial to analyse their politico-administrative transition.[6] This model illustrates the unidirectional transition from traditional agricultural societies to modern industrial ones and highlights the relationship between administrative behaviour and ecological factors in transitional societies. This is depicted in Figure 13.

Features of the Agraria

In the detailed analysis of the agraria-industria model, Riggs outlines the main characteristics of an agrarian society. Agrarian societies rely predominantly on agriculture for their economic sustenance. This dependence makes them somewhat backward compared to industrial societies, although they can still utilise modern agricultural tools. Social mobility is limited, and status is often determined by factors other than merit. Additionally, religion and politics play significant roles in agrarian societies, influencing people's behaviour and decision-making. Overall, the agrarian way of life is characterised by simplicity and austerity, in contrast to the pursuit of wealth in industrial societies.[7]

Features of industria

Riggs outlines the features of an industria as contrasting with those of an agrarian society. These features include the organisation of citizens into associations for specific purposes, a high degree of social mobility due to a skilled workforce, a fully developed occupational system and a religious society where religion does not control behaviour.

Transitia model

Riggs developed the transitia model as an alternative to the agraria–industria model, recognising that societies evolve and never remain in one stage permanently. He described agrarian societies as self-contained, with a belief in divine authority and communal or dynastic organisations, while industria societies

6 F. W. Riggs, *The Ecology of Public Administration* (New Delhi: Indian Institute of Public Administration, 1961).

7 F. W. Riggs, *Administration in Developing Countries* (Boston, MA: Houghton Mifflin, 1964).

had developed market economies, secular authority, and individualism. Riggs later introduced the Prismatic–Fused–Refracted Model.[8]

Shortcomings in the agraria-industria model

The agraria-industria model, developed by Riggs, had inherent shortcomings. First, it lacked a framework to study transitional societies. Additionally, it focused more on the ecology of the administrative system instead of the administrative system itself and failed to explain the administrative setup of new nations. The model also assumed unidirectional movement and did not address the coexistence of agrarian and industrial aspects within societies. Riggs recognised these limitations and developed the more refined prismatic–fused–refracted model to address these gaps and continue his research on contextual determinants.

Prismatic-fused-refracted model

Riggs's prismatic-fused-refracted model categorises societies based on their developmental background. The model includes a structural-functional analysis, decisional, fused, refracted society and prismatic models. Riggs uses various terminologies and examples to illustrate these models and emphasises that societies can fall between fused and refracted. The prismatic societies are positioned between these extremes and exhibit a blend of traditional and modern characteristics.

Ecological approach and development administration: The prismatic model

In the context of development administration, the prismatic model is characterised by heterogeneity, formalism, and overlapping. This Model challenges the traditional bureaucratic approach by advocating for position classification that combines duties, status, achievement and ascription. In contrast to diffracted formalised bureaucratic models, a prismatic society does not adhere to the principle of 'equal pay for equal work'; instead, officers' salaries are assigned to their rank rather than their work or achievements. This approach emphasises the importance of considering duties and status, achievement and ascription in position classification within development administration. Riggs gave two sub-models: the Sala model, which is a prismatic system, and the Bazaar-canteen model or the 'economic sub-system', which describes how the local officials and markets influence the common man.

---

8 F. W. Riggs, 'The "Sala" Model: An Ecological Approach to the Study of Comparative Administration', *Philippine Journal of Public Administration*, 6.1 (1962), 3–16.

Critical assessment of the Riggsian models

Critics argue that Riggsian models need to be more focused by reducing the scope in time and space to be empirically successful. They suggest the need for a middle-rung theory to restrict the set of relationships rather than analysing the entire social system as Riggs did. Some scholars criticise Riggs for overlooking different social structures, ignoring overlapping characteristics, and imbuing Western ideology into his work. However, Riggs' contributions to understanding development in fused and prismatic societies are still valued as he became a pioneer in understanding comparative public administration.

Antidevelopment (ADT) theory: A precursor to good governance

The years from the 1940s to the 1980s are called the 'Age of Development'. This epoch is coming to an end. The time is ripe to write its obituary according to Wolfgang Sachs.[9] In the process of emulation, the light of development got lost and darker because inequality and injustice did not reduce. The undeveloped and the developing countries were being fed by the rich developed countries with their own selfish interests of market and economic exploitation. The world embarked on a path of development following the aftermath of the world wars. If one divides the world into North and the South, it is found that the North is 'Westernised' and 'developed' and tried to dominate the South – the new as well as old nations, which were extremely poor, undeveloped and had no planning. The South looked towards the West and copied the Western developmental models because they were attracted by the Western way of life. In 1949, the President of the United States declared the Southern Hemisphere as the 'underdeveloped areas'. The United States seized the opportunity to be a 'beacon on the hill' amidst the collapse of European colonial powers, launching the concept of 'development' with a mix of generosity, bribery and oppression.

Dependency theory

The points to be considered before addressing dependency theory are:

- Developing countries have depended on the Western world, a concept arising in the 1960s from Prebisch's development theory.

---

9 Wolfgang Sachs, *The Development Dictionary: A Guide to Knowledge as Power* (London: Zed Books, 2010).

- Western-aided development processes disrupt the political, cultural and socioeconomic ecosystems of developing nations, pushing them towards industrialisation, profit-seeking, and resource exploitation.
- Developing nations often fall into debt traps in the name of development.
- Industrialisation and development displace people, causing social changes.

Dependency Theory has two basic assumptions.

- Core and periphery: Core countries are developed economies with high living standards and significant power, benefiting more than the poorer periphery countries, which focus on basic infrastructure and services.
- Condition of development: Core countries reinforce their capitalist order, exploiting periphery countries under the guise of development. Periphery countries struggle with structural weaknesses and external pressures, leading to 'dependency.'

Since the 1960s, global economic balances have shifted. Emerging markets like North America, Taiwan, Singapore, and the BRICS countries have risen.[10] Africa has seen 100 per cent economic growth, and India has become a large emerging economy.

Alternate method of economic development

The alternative methods of economic development for periphery countries give the idea of 'decoupling' from the capitalist system in favour of a 'socialist' mode of production. The relevance of the dependency theory in the context of contemporary global economics is in question because of its diminished applicability as the emerging markets success. The key points that emerge out of this claim are:

Dependence on core countries

Despite economic advancements in some periphery countries, a significant disparity remains between core and periphery nations in terms of income levels and living standards. Core countries continue to dominate international monetary agencies, reinforcing economic imbalances.

10 Brazil, Russia, India, China and South Africa are the member nations of the association BRICS, which was formed for newly advanced emerging economies. South Africa was made a member in 2010.

Economic conditions of periphery countries

Periphery nations remain economically weaker and rely on core countries for capital and technology. Developed nations maintain dominance through stringent import quality standards, hindering periphery countries' trade capabilities.

Breaking the dependency cycle

Some periphery countries have managed to overcome economic dependence through strategies such as import substitution, assistance from core countries, and the impact of slowed American productivity. Notably, India and China have emerged as strong economic powers due to well-directed policies. The severe environmental impact of industrialisation and the overexploitation of natural resources stress the unsustainability of current consumption patterns. While the concept of 'development' persists, the dependency theory has become obsolete as it fails to provide viable alternative development strategies. The economic landscape has evolved, necessitating new approaches to address the persistent issues facing developing countries.

Anti-development thesis (ADT)

The ADT theorists argue that development planning began post-World War II with the Marshall Plan to reconstruct war-ravaged Western nations. The same development aid model was applied to non-Western underdeveloped countries. Emerging from critiques of dependency theory in the late 1970s and 1980s, the ADT was later framed as good governance when international agencies introduced sustainable development. The ADT arose from the realisation that Western efforts often resulted in counter-development.

Key milestones in ADT evolution

1. United Nations resolution: Established a project for a unified approach to development integrating economic and social policies. Issues like environment, population explosion, hunger and women's empowerment became important, planting seeds of anti-development.
2. Declaration of Cocoyoc, 1974: Asserted that development should focus on human well-being rather than material growth.
3. 1975 Dag Hammarskjöld Report on Development and International Cooperation: Suggested a human-centred approach, termed 'another development'.
4. Johan Galtung: Advocated for the development focused on people, not material gains.

5. UNESCO: Promoted 'endogenous development', a holistic process involving all aspects of life and indigenous initiatives.[11]
6. International Labour Organisation (ILO), 1976: Proposed the 'basic needs approach' to achieve minimum living standards, emphasising direct need fulfilment over process-driven development.
7. Post-Development Age: The 1980s, known as the lost decade for development, shifted focus to technical and socio-political redevelopment, now termed sustainable development. However, some scholars criticise it for not supporting diverse natural and social life. In 1990, UNDP published its first human development report.
8. The economic man: Economists created the concept of the economic man to sustain markets. This construct facilitated economic society's emergence but was marked by historical violence and disruption.
9. New Commons: Gustavo Esteva, in *The Development Dictionary*, described resistance to economic encroachment as a creative reconstitution. New commons emerged in post-war social interactions, revitalising communities and cultures destroyed by industrial economic systems.[12]
10. Regeneration by the New Commons: New commons disengage from market logic to regenerate facilities autonomously. Key tenets include:
    - Education: Emphasising knowledge sharing outside formal diplomas.
    - Health: Reorganising health as autonomous environmental coping, reviving traditional healing with minimal external aid.
    - Production and food: Regenerating lives and lands through practices like intercropping, outside market failures.
    - Relocating: Assisting jobless people in relocation efforts.

    These efforts strengthen community interactions, countering the economic principle of exchange, calling for political action to view individuals beyond the 'economic man' concept.
11. Post-Development (Anti-Development): Thinkers like Wolfgang Sachs, James Ferguson, Arturo Escobar, Majid Rahnema and Serge Latouche, among others, proposed rejecting Western development strategies, viewing them as exploitative. They argued these strategies enriched only a few, leaving the majority without basic benefits and services. They advocated for a shift from traditional development models to more inclusive, equitable approaches.

---

11 Aram Ziai, 'Cord 2003', <http://www.globalpolitics.cz> [accessed 11 November 2017].

12 *Development Studies: A Reader*, ed. by Corbridge (London: Edward Arnold, 1995).

Alternatives to development by antidevelopment/ post-development theorists

The ADT theorists did not leave their thesis at the criticism of the developmental strategies but also suggested alternatives to it which are considered as a precursor to good governance. The decade of 1990 to 2000 saw globalisation, liberalisation and privatisation around the globe which led to the formulation of the MDGs and later the SDGs. The ADT theorists called for an action by all the 'good people' everywhere to get together and come up with solutions instead of just aping the West by implementing their Model without any modifications to suit their indigenous environment and a bottoms-up approach. It was the call for dead ministered development which was later known as participatory development and sustainable development. The alternative methods that they suggested are described further and depicted in Figure 14.

1. Participative action: The ADT theorists proposed participatory development where the state must take the responsibility but restrain from manipulation. Mostly, the participants do not even understand that they are being manipulated because they are directed by centres outside their control.
   - Resources: The local bodies must be given resources so that they can govern with an understanding of local needs and expectations of the people. The new modern nations were grappling with their new-found freedom to deliver public services with effectiveness and efficiency while they had to deal with their economies.
   - Decentralisation: Decentralised governance will strengthen the institutional framework resulting in catering to the political demands and becoming more responsive to global economic changes. Small-scale enterprises will also participate on the economic front when given the opportunities, for example, the Global Investors Summit in the state of

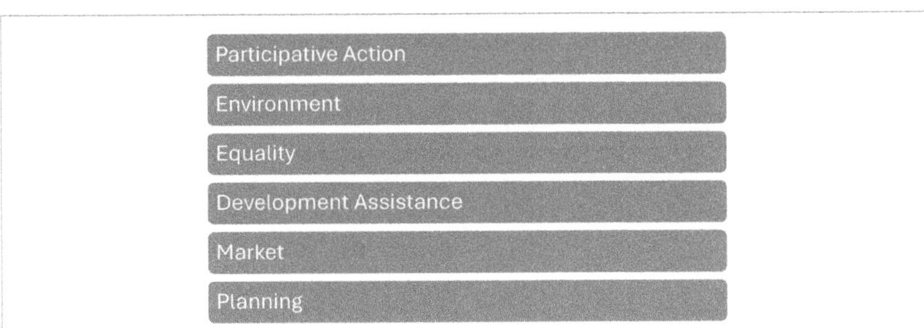

Figure 14 Alternate methods given by the ADT

Madhya Pradesh in India where many global companies came to invest in manufacturing units.

- Popular participatory approach: Popular participatory approach is defined as 'the organised efforts to increase control over resources and movements'.[13] A genuine dialogue must be initiated between the government agencies and the affected people by giving them full opportunity to participate.

2. The will of the people: The will of the people manifests when citizens organise themselves, ensuring their expectations are heard, leading to humanitarian and moral outcomes. This supports a human-centred approach. Participatory action research fosters participatory development by mobilising citizens to address their own issues realistically. James Ferguson highlighted that in failed development projects, the focus should be on action rather than blame. Developing countries struggle to formulate effective socio-political plans, relying on NGOs to engage 'change agents' for awareness.[24] Ultimately, unbiased citizen involvement is crucial for meaningful societal change, even though small organisational efforts.
3. Environment: ADT theorists argue that development projects often result in environmental damage and resource depletion, contrary to their intended outcomes. They critique how the concept of development is framed by developed nations, emphasising issues like pollution, acid rain, and resource exploitation raised since the 1972 UN Conference on the Human Environment in Stockholm. The 1987 Brundtland report underscored the interplay between development and environmental sustainability, highlighting industrialisation's destructive impact. ADT proponents advocate for sustainable development, asserting that poverty reduction cannot be achieved sustainably without preserving natural resources – an essential critique of traditional development ideologies focused solely on economic growth.
4. Equality: The ADT theorists stated that the problem of inequality is the problem of the world's rich rather than the poor. The developed nations need to check their over- consumption which requires them to exploit the developing countries to produce more. The developed nations give aid by way of developmental planning projects, but the developing countries fall into the debt trap. The condition worsens and inequality also.

---

13 M. Wolfe, *Participation: A Conceptual Framework* (Geneva: United Nations Research Institute for Social Development, 1976).

5. Development assistance: The developed nations give development assistance with conditions and threats of sanctions to the developing countries in an institutional way. Poor quality and discarded products from the junkyards in the name of military aid and seed grains to control the food supplies are sent off to the third world countries. The ADT theorists promoted the concept of help for self-help so that the poor developing countries can learn to be independent. The developed countries must not attempt to control the levels of people in the developing countries and must help them to follow their own way of life. They must acknowledge equality of various cultures.
6. Market: Economies must integrate themselves into the national and international markets to be considered as a developed nation. Markets are an integral part of any development theory or policy. A question was raised by the ADT theorists if the market must be contained by the society, or it must be left free to regulate the society in totality.
7. Planning: Planning process, methodology and procedures have been a part of having effective development administration for the third world countries or the Southern Hemisphere. The later eighteenth century linked development and planning intrinsically that led to the western modernity and carried the historical burden when applied to the developing nations. The only thing that it has resulted in has been underdevelopment.

The ADT theorists' critique Western perspectives that view the non-Western world through a lens of poverty, overlooking rich cultural and social systems and abundant natural resources. They argue that development planning must redefine categories and techniques. Western theories emphasise increased production as development, dismissing subsistence economies as inefficient. ADT highlights industrialisation's unintended environmental harm, from pollution to cultural erosion, affecting even the service sector. They advocate a people-centric approach, rejecting Western development models in favour of indigenous methods to preserve cultural integrity. They propose decentralised, citizen-engaged policymaking tailored to each nation's economic and sociocultural context, opposing a universalised development approach. This paradigm shift aims to empower local communities and safeguard their traditional ways against Western-centric development ideologies.

Shortcomings of the Anti-Development Thesis (ADT)

In the late 1970s and 1980s, the ADT gained momentum, critiquing Western development models. However, it failed to provide effective alternatives, focusing more on criticism than proposing definitive models. Key shortcomings include:

- Oversimplistic rejection: Dismissing Western development without considering historical context is overly simplistic. Post-colonial third world countries accepted Western development terms to survive and grow, making outright rejection unrealistic.
- Acknowledging empowerment: Criticising modernity and science ignores their benefits, like individual rights and modern medicine, which improved health and life expectancy. ADT overlooks the empowerment and assistance Western development provided, evidenced by a 40 to 50 per cent increase in the Human Development Index of developing countries.[14]
- Vague alternatives: ADT's preference for traditional methods could lead to indigenous imperialism, concentrating power and wealth among elites, causing unequal distribution. Western development strategies have benefited common masses in developing countries, suggesting these nations may struggle without Western inputs, risking human rights violations.
- Globalisation issues: ADT fails to explain how third world countries would handle globalisation using only indigenous methods.
- Rigid Critique: ADT theorists focused on critique without offering practical implementation suggestions, lacking a balanced perspective.

ADT: A precursor to good governance

To sum up, the ADT theorists were able to highlight some important points and issues which were subsequently considered and adopted by international agencies, although they had to be worked upon and modified to form them into workable solutions. Decolonisation is important to push forward development administration in the developing countries. The term' good governance' rendered the ADT theory its precursor. This account sums up the theoretical base of development administration from an international perspective. The present status of development administration can be viewed in its transformation into sustainable development and good governance as both the concepts aim at the same goals which development administration has.

Conclusion

In conclusion, the integration of development administration theories into global governance frameworks demonstrates how administrative systems can both support and challenge sustainable development efforts. From an international

14 Human Development Report 2023–24, <https://hdr.undp.org/content/human-development-report-2023-24> [accessed 20 September 2024].

perspective, development administration is pivotal in creating structures that foster economic growth, social equity, and environmental sustainability. For example, Karl Marx's critique of capitalist-driven development remains relevant in international development debates today, especially in regions where global financial systems have perpetuated inequalities. The growing influence of anti-development theories highlights how the neoliberal approach to development, characterised by market-driven reforms, can fail to meet the needs of marginalised communities, particularly in Sub-Saharan Africa and Latin America. The ongoing calls for good governance frameworks echo this critique, as they stress transparency, accountability, and participatory governance – an area where both developed and developing nations continue to grapple with entrenched bureaucracies and corruption.

The bureaucratic model, as implemented in nations like Germany and France, provides a structured approach to development administration but often faces criticism for being overly rigid and slow to adapt to evolving global challenges. Conversely, the behavioural model, embraced in countries such as Canada and Australia, highlights the role of individual actors and their decision-making processes in fostering more dynamic and responsive systems. The involvement of multiple stakeholders in shaping policy outcomes, from citizens to administrators, is crucial in aligning national development goals with the broader objectives of the Sustainable Development Goals (SDGs).

The Weidner and Riggsian models offer essential lessons for developing countries striving to build robust administrative capacities. The success of Singapore's development administration, for instance, reflects the need for strong institutional frameworks that adapt to environmental and societal changes. Riggs's ecological approach is particularly relevant in Southeast Asia, where diverse socio-cultural factors and complex administrative ecosystems interact with development policies, especially in addressing the challenges of urbanisation and environmental degradation.

Finally, the rise of sustainable development governance emphasises the need for development administration to shift from traditional growth-focused models to those that prioritise sustainability. Countries like Norway and Sweden have successfully merged development administration with sustainable practices, setting global benchmarks in areas like renewable energy and ocean conservation. These examples underline the importance of tailoring administrative systems to national contexts while adhering to international standards of governance, equity, and environmental stewardship. Thus, international development administration must continue evolving to meet the challenges posed by economic, social, and environmental shifts while maintaining a focus on governance that ensures inclusive and sustainable progress.

CHAPTER 3

The Rise of Sustainable Development Administration

A Paradigm Shift

Introduction

In the 1950s and 1960s, the purpose of development administration in modern nations was to reconstruct their war-torn areas and simultaneously raise the levels of socioeconomic status. The world was facing extreme poverty, low economic growth and a collapse of social values. European nations had been war-ravaged, while the southern hemisphere had just been freed from the imperialist forces. The West began to formulate development models and found success in their countries. In contrast, the least developed economies in the southern hemisphere were desperately looking for a new methodology or model for their economies.

In the absence of any development models specific to their countries, they began adopting the Western development models, which failed miserably in the underdeveloped countries. The problem was compounded by developmental aid because the West gave it to underdeveloped countries in the name of development, but the latter got trapped. They were unable to get out of the rut of receiving aid, whether financial or technical and eventually, the situation became complex and worse. They became dependent and under debt, forgetting their own culture and the fact that they were economically surviving before the imperialists took over. They forgot how to be self-reliant and economically independent.

By the 1970s, development administration models of the West had failed in developing countries. Then came the period when Western scholars, post-realization of the failure of Western development models, began formulating development models specifically for developing nations. The developing nations also had matured by then and tried their methods. The situation still did not improve as they could not find ways to become economically independent and sustainable. The international agencies took cognizance of this significant issue and, by the new millennium, had formulated Millennium Development Goals (MDGs). International guidelines were made to achieve public service delivery standards, and country-specific partnerships were drawn to guide developing countries out of their misery. MDGs saw some success, and they completed their tenure in 2015, and the Sustainable Development Goals (SDGs) Agenda 2030 was introduced.

Transformation of development administration into sustainable development

The paradigm shift from development administration to sustainable development has made the thinkers and policy-makers reinvent this discipline. The thrust of development can never be taken away. However, our approach to achieving it can be modified and upgraded according to the dynamically changing economic and social environment. Contrary to what some scholars say, development administration has lost its importance in the context of globalisation, technical innovation and awareness of ecological issues,. It is now the responsibility of sustainable development administration to keep up with the changing times. The end user of all developmental schemes in sustainable development administration or governance is the same as in development administration. However, the processes to successfully implement the development project schemes and the objectives, goals, and procedures for implementation have transformed towards sustainability.

The 1950s saw the emergence of development administration, and as mentioned earlier, multiple theories, approaches, strategies, policies, models and studies were conducted to bring about effective and efficient development. Although the results focused on growth, the new nations could not make much progress or do so sustainably. Sustainable development administration can be viewed as a paradigm shift and not its demise, as some would say. The transformation of procedures is the evolution and growth of the discipline, not its losing relevance and importance. Simply put, sustainable development administration would have never reached where it is if development administration had not happened, and the former can be considered as the next phase of the discipline.

There came a paradigm shift in development administration, which came to be guided by the SDG Agenda and has been formulated to remove significant obstacles in developing countries' development. Development administration has simply transformed into sustainable development administration based on the demands of the situation. The term 'sustainable' implies the exact results the development administration has aimed for since its inception but failed to attain. Governments worldwide were concerned with the failure of development administration in developing countries. The low status of poverty and economic growth in these developing countries meant low spending power for consumers and, consequently, a lack of market for the developed countries to sell their goods. The developed nations were affected by the poor economic conditions of the Southern Hemisphere as it is said that 'If my neighbour is poor, it affects me too'.

The issues to be studied and carefully considered relate to the emerging tripartite philosophy of 'global village', 'global citizenship' and 'refounding' of development administration as well as its separation from politics altogether, all of which redefines the objectives and activities of the governments. Scholars are continuously working to define the paradigm shift in development administration and make sustainable development a new area of focus.

Backtracking sustainable development

The following is a timeline of how the concept of sustainable development evolved over the years.

1713 Hans Carl Von Carlowitz wrote on sustainable forestry in his publication '*nachhaltendenutzung*' (sustainable use).

1972 The term 'sustainable' is concerned with the ecosystems. Fisheries managers used the 'maximum sustainable yield' concept to define the maximum fish catch per year with a stable fish population. The United Nations Conference on the Human Environment was held in Stockholm, and the main theme was to discuss the issue of maintaining 'sustainability in economic growth and development'. The Club of Rome published a book entitled *The Limits to Growth*, making a crucial and relevant point that continued economic growth on the prevailing economic patterns would conflict with the Earth's finite resources, leading to a future overshoot and collapse.[1]

1980 The *World Conservation Strategy: Living Resource Conservation for Sustainable Development* was published to help advance sustainable development by conserving living resources. The concept of sustainable development was initially given in the *Brundtland Report* by the United Nations Environment Commission, chaired by Gro Harlem Brundtland, the Prime Minister of Norway.[2] The concept began with the environment as its basic premise. Sustainable development was defined as 'development that meets the needs of the present without compromising the ability of future generations to meet their own needs'. It was popularly known as the Brundtland Report, with which two main concepts are fundamentally tied to the process of sustainable management of the Earth's resources:

- The basic needs of humanity, such as food, clothing, shelter and jobs, must be met. This involves paying attention to the largely unmet needs of the world's poor, which should be given overriding priority.

1 Donella H. Meadows et al., *The Limits to Growth: A Report for the Club of Rome's Project on the Predicament of Mankind, Part 1* (New York: Universe Books, 1972).

2 International Union for Conservation of Nature and Natural Resources and United Nations Environment Programme, *World Conservation Strategy: Living Resource Conservation for Sustainable Development* (Gland: IUCN, 1980).

- The limits to development are not absolute but are imposed by present states of technology and social organisation, their impacts on environmental resources, and the biosphere's ability to absorb the effects of human activities. Still, technological and social organisation can be managed and improved to create a new era of economic growth.[3]

1987 The World Commission on Environment and Development was established in 1983 and published its report 'Our Common Future', widely recognised as 'The Brundlandt Report', in 1987. The pursuit of sustainable development unequivocally demands the following:

- A political system that secures effective citizen participation in decision-making
- An economic system that can generate surpluses and technical knowledge on a self-reliant and sustained basis
- A social system that provides solutions for the tensions arising from disharmonious development
- A production system that respects the obligation to preserve the ecological base for development
- A technological system that can search continuously for new solutions
- An international system that fosters sustainable patterns of trade and finance
- An administrative system that is flexible and has the capacity for self-correction

1992 The 'intergenerational' concept of sustainable development was adopted by most member states at the UN Conference on Environment and Development held in Rio. The 'Precautionary Principle' adopted at the resolution stated the following:

- No environmental action should be taken which is not reversible.
- Designers should use the best scientific knowledge available.
- Scientists must develop environmental knowledge.
- Ignorance was no defence under international law for ecological damage.[4]

2000 The Earth Charter propagated the participatory process involving global Civil Society.

---

3 Don Hinrichsen, *Our Common Future: A Reader's Guide* (Washington, DC: Earthscan, 1987).

4 Report of the United Nations Conference on Environment and Development, Rio de Janeiro, 3-14 June 1992, Annex 1: Rio Declaration on Environment and Development, UN Doc. A/CONF.151/26.

2002 The 'intergenerational' concept of sustainable development was widely adopted at Rio's United Nations Conference on Environment and Development.

2002 The Johannesburg World Summit on Sustainable Development was held and it introduced the concept of 'sustainable consumption and production', leading to several international agreements. The fundamental principle was establishing a link between productivity, resource use and population levels. The agreement was about the following:

- Ensuring that economic growth does not cause environmental pollution at global and regional levels
- Improving efficiency in resource use
- Examining the whole life cycle of a product
- Giving consumers more information on products and services
- Exploiting taxation and regulation to stimulate innovation in clean technologies

2012 In its outcome document, 'The Future We Want',[5] in the three-part vision of sustainable development – economic development, social development and environmental protection – on the twentieth anniversary of the Rio Summit, it was stated 'We also reaffirm the need to achieve Sustainable Development' to:

- Promote sustained, inclusive and equitable economic growth
- Create more significant opportunities for all
- Reduce inequalities
- Raise basic standards of living
- Foster equitable social development and inclusion
- Promote integrated and sustainable management of natural resources and ecosystems that supports, among other things, economic, social and human development while facilitating ecosystem conservation, regeneration, restoration and resilience in the face of new and emerging challenges.

The SDGs should address and incorporate all three dimensions and their interlinkages in a balanced way. They must be action-oriented, concise and easy to communicate: limited in number, global and universally applicable to all

---

5 Report of the United Nations Conference on Sustainable Development, Rio De Janeiro 20-22 June 2012, UN Doc. A/CONF.216/16.

countries while considering different national realities, capacities and levels of development and respecting national policies and priorities.

The concept of sustainable development

The concept of sustainable development has two main features attached to it:

1. Normative outlook: After a careful study of the evolution of sustainable development, it is observed that the subject has taken the shape of a practical model. The three-way normative framework development includes economic development, social inclusion, and environmental sustainability. The basic concept of sustainable development is the same even today. It is also an analytical field that aims to explain and predict complex national systems.[6]
2. Complexity: Besides being a normative and ethical concept, sustainable development is a science of complex systems. The economy is a system of millions of individuals and businesses bound together in markets, contracts, laws, public services and regulations. A growing economy is more than a sum of its businesses and workers. Even a tiny change, like the failure of one business, can cause a cascading financial panic and a global downturn like the 2008 crisis, which occurred due to the failure of Lehman Brothers Investment Bank.

Sustainable development encompasses a range of critical issues, which are outlined below and best understood through the work of Jeffrey Sachs.

Interacting systems of sustainable development

According to Jeffrey Sachs, sustainable development operates through the interaction of three fundamental systems: economic development, environmental sustainability and social inclusion. These systems must function harmoniously to achieve long-term well-being. Sachs stresses that the integration of these systems is crucial for achieving sustainable progress on a global scale. Sachs gives four interacting systems of sustainable development in his most acclaimed book, *The Age of Sustainable Development*.

- Economic: Economic systems are central to sustainable development, shaping how resources are allocated, wealth is generated, and poverty is addressed.

6 Jeffrey Sachs, *The Age of Sustainable Development* (New York: Columbia University Press, 2015).

Sachs emphasises that economic growth must be inclusive, lifting people from poverty while maintaining ecological balance. This involves increasing incomes and ensuring green growth, meaning it does not deplete natural resources or cause environmental degradation. The governments must address global concerns like inequality, access to healthcare and infrastructure development to ensure long-term prosperity. Economic policies must align with environmental sustainability and social inclusion to be effective.

- Social: The social system in sustainable development focuses on the well-being and interactions within communities, emphasizing equality and inclusion. According to Sachs, trust, ethics and social networks are critical to building a fair and resilient society. Social inclusion means ensuring that marginalised populations, including those in poverty, women, and minority groups, have access to essential services such as healthcare, education and housing. Strengthening social support networks in physical communities or online fosters cooperation and reduces inequality. Without social cohesion and justice, sustainable development cannot be fully realised.
- Environmental: Sachs highlights the critical role of environmental sustainability in maintaining the planet's life-support systems, including the atmosphere, water cycles and ecosystems. Responsibly managing natural resources is essential for avoiding catastrophic consequences like climate change, biodiversity loss, and ecosystem collapse. Sustainable development must align human activities with the planet's ecological limits, ensuring that development meets present needs without compromising future generations. Protecting forests, oceans and wildlife while addressing pollution and greenhouse gas emissions is crucial to safeguarding Earth's natural processes and promoting a balance between development and conservation.
- Governance: Governance plays a pivotal role in sustainable development by shaping policies, laws, and institutions that guide the public and private sectors. Sachs underscores that effective governance ensures accountability, transparency, and ethical decision-making from local governments to multinational corporations. Governance also includes managing public goods, such as education, health and infrastructure, while regulating markets to prevent corruption and ensure equitable resource distribution. Businesses and governments must collaborate to create frameworks that incentivise sustainable practices, ensuring that long-term economic, social, and environmental goals are achieved.

All the aforementioned interacting systems are complex and there can be no one solution. What follows is a closer look at the additional factor of technological progress and innovation.

Technological change

Technological progress and innovation are pivotal in driving sustainable development but are not without challenges. Innovation plays a dual role as both an enabler of economic growth and a contributor to environmental and social issues. While it has the potential to benefit society, its unintended consequences must be anticipated beforehand so that action to resolve them can be planned. Technology can have far-reaching positive impacts by focusing on human-centred goals. The private sector, governments and researchers must collaborate to harness innovation for the betterment of society, ensuring that technological advances support both the environment and marginalised communities. The three aspects of technological challenges are given below.

- Enablers: Technological innovation is a critical enabler of long-term economic growth, driving productivity, efficiency and sustainability. It has revolutionised industries and sectors, from agriculture to healthcare, helping to solve complex challenges. When aligned with sustainable development goals, technology can promote greener energy, more efficient resource use and innovative solutions to combat poverty. Examples include using renewable energy technologies like solar and wind power to reduce carbon footprints. Societies can harness the power of technology to achieve broader social, economic, and environmental objectives, laying the foundation for long-term prosperity by investing in research and development.
- Adverse Side-effects: Despite the significant benefits of technological advances, they often bring unintended negative consequences. For example, while electric trains reduce carbon emissions compared to coal-powered ones, they may still rely on electricity from non-renewable sources, shifting rather than eliminating environmental harm. The Industrial Revolution, fuelled by coal, led to economic expansion and laid the groundwork for the current environmental crisis through massive pollution. Technological innovations must be critically assessed to mitigate their adverse effects, ensuring that progress does not come at the expense of future generations and the environment.
- The Goal: The ultimate goal of technological advancement should be to address human-centred challenges, particularly for marginalised populations and the environment. While innovations may drive profits for corporations and researchers, their value lies in solving global problems like poverty, inequality and climate change. The private sector and government institutions must prioritise research and development efforts with broad social impacts. Collaborative efforts can direct technological progress toward equitable growth, ensuring that innovation benefits everyone, especially

those who are most vulnerable, while safeguarding the planet's natural resources for future.

Ethical nature of sustainable development

Sustainable development is inherently ethical, focusing on what constitutes a just and good society. It emphasises a holistic approach, recognizing that achieving a high average income is insufficient. A just society requires equitable distribution of resources, with income distribution crucial to long-term social and environmental sustainability.

- Extreme poverty: Sustainable development prioritises eradicating extreme poverty, ensuring that no individual is deprived of basic human needs such as food, clean water, shelter, and healthcare. Dealing with extreme poverty is fundamental to achieving broader economic and social goals, as it directly impacts human well-being, equality and global stability.
- Inequality: Inequality disrupts social cohesion and undermines sustainable development efforts. Reducing income disparity is crucial to fostering social equity and long-term economic growth. A society with vast disparities in wealth and resources is likely to experience social unrest and face challenges in achieving holistic and inclusive development.
- Social mobility: Social mobility ensures that individuals can improve their socio-economic status through education, employment, and access to resources. Sustainable development advocates for systems that allow for upward mobility, ensuring that all members of society can contribute to and benefit from economic progress, regardless of their starting point.
- Discrimination: Combating discrimination is essential for creating an inclusive and just society. Sustainable development calls for eliminating biases based on race, gender, religion or socio-economic status. Ensuring equal rights and opportunities for all fosters a culture where diversity thrives, leading to innovation, fairness and long-term prosperity.
- Social cohesion: Social cohesion refers to strengthening relationships and solidarity among community members. A cohesive society, where individuals feel connected and supported, is better equipped to work collectively toward sustainable goals. Social bonds promote trust, cooperation, and shared responsibility, which are essential for long-term development.
- Moral code: A moral code provides the ethical framework that guides societal decision-making and behaviour. Sustainable development means fostering values like justice, equity, and stewardship of natural resources. A robust moral foundation ensures that economic and social policies are aligned with the long-term well-being of people and the planet.

Thus, the ethical framework of sustainable development ensures that economic growth, social justice, and environmental stewardship are harmonious, promoting a society that values equity, inclusiveness and long-term well-being.

Environment

A good society must prioritise environmental conservation to ensure that future generations inherit a planet with adequate resources and ecological balance. Sustainable development is inherently intergenerational, as it requires us to protect and manage natural resources in a way that meets present needs without compromising the ability of future generations to meet theirs. Critical environmental systems – such as water, biodiversity, oceans and rainforests – are essential to sustaining life on Earth. The degradation of these systems through deforestation, pollution, and over-extraction threatens the environment and the social and economic well-being of communities worldwide. For instance, the loss of biodiversity weakens ecosystems, reduces agricultural productivity and diminishes the availability of medicinal resources. Likewise, the health of oceans and rainforests is crucial for regulating climate, providing livelihoods, and maintaining ecological diversity. Addressing these environmental challenges through sustainable policies and practices is vital for ensuring a just and prosperous future for all generations.

Good governance

Good governance is a fundamental pillar of sustainable development, as it ensures that political, legal, and institutional systems function transparently, efficiently, and equitably. Effective governance upholds the rule of law, which fosters a sense of security and stability among citizens. Without the rule of law, societies are vulnerable to corruption, lawlessness and the abuse of power by untrustworthy politicians. These issues undermine social cohesion, deter investment, limit economic growth, and perpetuate inequality. Fair and impartial government services are essential to building public trust in governance systems. When people trust their leaders and institutions, they are more likely to engage in civic processes and support policies that promote long-term sustainable development. Transparent decision-making, accountability, and inclusive governance empower communities, reduce corruption, and ensure that resources are distributed equitably. Good governance is the foundation for sustainable economic growth, social inclusion, and environmental protection, providing a resilient and prosperous society.

To sum up, from the points mentioned above, a workable definition of sustainable development can be: A good society is not only an economically prosperous society (with high per capita income) but also one that is socially inclusive, environmentally sustainable and well governed.[7] This view has been endorsed by the SDGs adopted by the United Nations member states.

Principles of sustainable development

The concept of sustainable development and its implementation is based on certain principles to develop and harmonise the SDGs to achieve worldwide developmental progress. They provide guidelines to identify priorities to design specific goals and tasks in a coordinated manner. The relevant bodies of the United Nations and the European Union have framed, clarified and adopted the basic principles at the highest levels. The following are the pertinent principles in forming sustainable development strategies, which are pretty similar to the conceptualization of sustainable development given by Jeffrey Sachs.

1. Holistic approach: A holistic approach emphasises the interconnectedness of systems, where changes in one element can influence others, even those far removed. Each system interacts with and impacts the broader environment. Tackling the local challenges requires considering local conditions and global trends to create effective solutions. Understanding these interconnections allows decision-makers to craft strategies that are resilient, sustainable and adaptive to the complexities of our world. This approach ensures that interventions are mindful of the ripple effects they may cause, ultimately promoting long-term, positive outcomes across different sectors and systems.
2. Intragenerational and intergenerational solidarity: Intragenerational and intergenerational solidarity is at the heart of sustainable development, emphasizing a balance between the needs of both current and future generations. Intragenerational solidarity ensures that today's development benefits all people equitably, addressing disparities in wealth, access to resources, and quality of life. It calls for the fair distribution of resources within and across societies, recognizing that poverty, inequality, and environmental degradation are global challenges that must be tackled collectively. On the other hand, intergenerational solidarity focuses on ensuring that the actions and decisions made today do not compromise the well-being of future generations. Sustainable development requires careful stewardship of natural resources, ecosystems, and economic systems, ensuring that future generations inherit a planet that can support their needs. This means mitigating environmental degradation, climate change and resource depletion

7 Ibid.

while promoting economic growth and social inclusion. The principle of solidarity underscores the interconnectedness of human well-being across time, reinforcing that a just and sustainable society considers both present and future needs equally.

3. Social justice: All individuals should have access to the resources and opportunities necessary for a dignified life while upholding fundamental human rights. Sustainable development recognises that equitable access to education, healthcare, housing, and economic opportunities is essential for fostering inclusive growth. Social justice ensures that development benefits all members of society by reducing inequality and ensuring that marginalised populations are not left behind. This principle supports creating systems that promote equal opportunities for acquiring skills and knowledge, empowering individuals to contribute to sustainable, resilient communities.
4. The principle of integration: A holistic approach is essential when formulating, evaluating, and implementing policies, plans, and programs. Economic, social and environmental considerations must be interlinked and treated as mutually reinforcing elements. This ensures that progress in one area does not come at the expense of another, promoting balanced and sustainable outcomes. Coordinating local, regional, and national activities is crucial to avoid conflicts and ensure alignment toward shared goals. Sustainable development fosters inclusive growth, social equity, and environmental protection, creating long-term, systemic solutions that benefit all stakeholders by integrating these dimensions.
5. The principle of using local resources: The principle of using local resources in sustainable development highlights the importance of fulfilling community needs through sustainable, locally sourced resources. By harnessing local resources, communities can promote economic development while preserving their unique environmental features and biodiversity. This principle also highlights the importance of safeguarding the man-made environment and protecting cultural heritage, recognizing that cultural and historical assets are integral to a community's identity and resilience. Sustainable development, therefore, requires a balance between utilizing local resources and protecting local ecosystems and cultural diversity for future generations.
6. The principle of management of resources: The principle of management of resources stresses the need to use resources within the environment's carrying capacity, ensuring they are not overexploited. Sustainable practices must be in place to preserve resources for future development. It is crucial

to recognise that natural resources, including biodiversity, are finite and integral to maintaining ecological balance. Conservation of biodiversity, which supports ecosystems and human life, should be prioritised to prevent the irreversible loss of species and habitats. Sustainable management ensures that current resource use does not deplete availability for future generations, supporting both immediate development needs and long-term environmental sustainability.

7. The principle of participation: Public participation in sustainable development highlights the importance of giving citizens access to information that affects their socio-economic and environmental well-being. Transparency allows individuals to engage in discussions and contribute to decision-making processes, ensuring policies reflect community needs. Educating people on sustainable development challenges, such as climate change and resource management, empowers them to understand the broader impact of their actions. Raising awareness about social, economic, and environmental issues enables communities to make informed decisions. Active public participation fosters a sense of ownership and responsibility, helping to create lasting, sustainable solutions for future generations.
8. The principle of 'social responsibility': Jeffrey Sachs' principle of social responsibility emphasises the interconnectedness of global challenges and the need for comprehensive, sustainable solutions. He advocates for environmental stewardship, urging responsible resource management and protection of ecosystems for future generations. Sachs also highlights social equity, stressing the importance of reducing poverty and inequality by ensuring that marginalised groups have access to education, healthcare and economic opportunities. His approach includes economic inclusivity, where economic growth benefits all segments of society. Additionally, Sachs calls for global cooperation, recognizing that international collaboration is essential to effectively deal with pressing issues like climate change, poverty, and sustainable development.
9. The principle of precaution and prevention: This principle is inherent to sustainable development, focusing on taking proactive measures to avoid environmental and human health harm. It stresses that when an action or policy poses potential risks, even if the full extent of the damage is uncertain, precautionary steps should be taken to prevent adverse outcomes. This principle encourages decision-makers to anticipate and mitigate risks before they materialise rather than waiting for conclusive evidence of harm. Prevention also includes minimizing environmental degradation

by promoting responsible resource management, reducing pollution and supporting innovations safeguarding ecosystems and public health.[8]

Hence, human activities must be planned and carried out according to the precautionary principle. There must be proactive action to prevent pollutants (like carbon emissions, e-waste, industrial waste) from adversely impacting the environment, seriously hampering the sustenance of natural systems and human health. Suppose it is not possible in some areas to do so. In that case, such activities must be reduced, and damage must be corrected and restored to its original condition as soon as possible.

10. The polluter pays principle: Prices must reflect the actual costs paid by society for activities involved in consumption and production and their impacts, including the costs of using natural resources. Those engaged in activities damaging or polluting the environment must pay for the damage caused to human health or the environment.[9] The polluter pays principle is a cornerstone of environmental law and sustainable development, holding that those responsible for environmental damage must bear the costs of managing and mitigating that harm. This principle aims to internalise the external costs of pollution, ensuring that polluters, rather than society or future generations, are financially accountable for their actions. It promotes adopting cleaner, more sustainable practices by incentivising businesses and individuals to reduce their environmental footprint. The polluter pays principle encourages responsible resource use, pollution control, and restoration efforts, contributing to long-term environmental protection and sustainable economic development.
11. The principle of equity: The equity principle has been given by the Multidisciplinary Digital Publishing Institute,[10] which could be subdivided into the following:
    - Intergenerational equity: It refers to the long-term or futurity aspect of sustainability as the concept not only aims to meet present human needs and aspirations but also includes the right of future generations to meet

8 Global Development Research Centre, 'The Precautionary Principle in Environmental Law', <https://www.gdrc.org/u-gov/precaution-7.html> [accessed 25 September 2024].

9 Tankönyvtár, 'Sustainable Development Resources',<https://www.tankonyvtar.hu> [accessed 25 September 2024].

10 MDPI, 'Sustainability Journal', <https://www.mdpi.com/journal/sustainability> [accessed 25 September 2024].

their needs and aspirations as we depend on natural resources to do the same. This principle consists of the requirement to keep within the environmental limits of the Earth. In this sense, 'We have not inherited the Earth from our parents; we have borrowed it from our children'.

- Intragenerational equity: It is similar to intergenerational equity but concerns the present generation. It refers to the realization of contemporary social equity, which is the right of every human being of the present generation to a decent quality of life.
- Geographical equity: Geographical equity or global responsibility refers to the need for cooperation worldwide and from the local to the international level in a spirit of 'shared but differentiated responsibility' in tackling sustainable issues oriented around acting locally and thinking globally.
- Procedural equity: It refers to democratic and participatory governance systems involving concerned stakeholders in decision-making. This is particularly essential because of the normative nature of sustainability.[11]
- Interspecies equity: It stands for environmental stewardship and refers to the survival of species on an equal basis to human survival. It highlights the critical importance of preserving ecosystem integrity and maintaining biodiversity for utilitarian reasons and because other species have the intrinsic right to survival.

12. The principle of dynamism: The dynamism principle states that sustainable development is a continuous process of directed sustainable change and not a fixed and predetermined end state. The societies and the environment in which they live incessantly interact and continue to evolve continuously. However, this appears to be like an excuse to exit from the escape route from the challenging task of achieving sustainability goals. Thus, sustainability must be strived for by building positive societal and political will and must be considered an uninterrupted search for a delicate equilibrium in a dynamic setting. Dynamism implies accepting its limitation of stability, working with innovation, and evolving new ways to adapt to changes voluntarily.

The principles mentioned above of sustainable development are broad-based and form the foundation upon which the international, national, state and local levels of governments and agencies build the SDGs. However,

11 Borgen Project, 'Poverty Reduction and Development', <https://www.borgenproject.org> [accessed 25 September 2024].

there are some limitations to these principles and can be accepted while developing SDGs:

- The principles have been stated generally, but they need further detailed application while being implemented.
- The ethical issues are difficult to quantify and its inculcation is a gradual process.

These limitations highlight the need for strong governance and global cooperation to implement sustainable development principles effectively.

SDG Agenda 2030

The 193-member United Nations General Assembly (UNGA) formally adopted the 2030 Agenda for Sustainable Development, along with a set of bold new global goals, which the Secretary General Ban Ki-Moon hailed as a universal, integrated and transformative vision for a better world on 25 September 2015 in New York. The UN Chief's address came before the assembly's formal adoption of the new framework, 'Transforming Our World: The 2030 Agenda for Sustainable Development', which comprises seventeen goals and 169 targets to wipe out poverty, fight inequality and tackle climatic issues over the next fifteen years.[12]

Transforming our world: The 2030 Agenda For Sustainable Development

The framework given in the 'Transforming Our World: The 2030 Agenda for Sustainable Development' begins with the preamble. The main points that were made in the preamble are given here in brief:

- The agenda is a plan of action for people, the planet and prosperity to strengthen universal peace and greater freedom.
- Eradicating poverty in all its forms and dimensions is the biggest global challenge and requirement of sustainable development.
- All countries and all stakeholders will implement this plan in collaboration.
- The motto of this agenda is to 'leave no one behind'.

---

12 SDG Cities, 'Sustainable Urban Development', <https://www.sdg-cities.org> [accessed 25 September 2024].

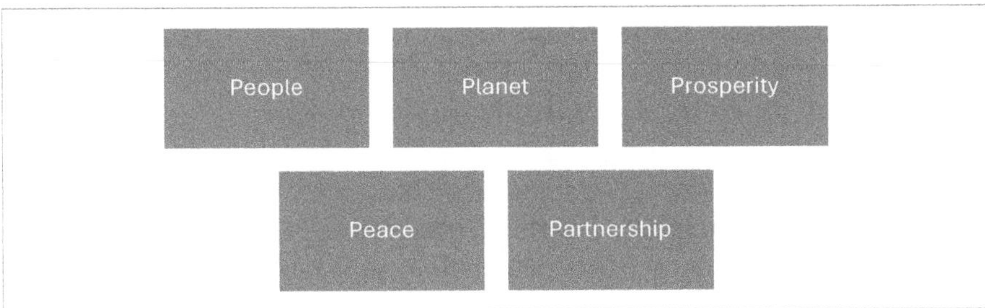

Figure 15 The five Ps

- The seventeen goals and 169 targets in the universal agenda seek to build on the MDGs and complete what they could not achieve.

The SDGs are integrated and indivisible and balance the three dimensions of sustainable development – the economic, the social and the environmental.

Importance for humanity and the planet: The five Ps

The goals and targets will stimulate action over the next fifteen years in areas of critical importance for humanity and the planet. Their focus will be on the five Ps depicted in Figure 15.

1. People: The primary objective is to eradicate poverty and hunger in all forms, ensuring that every individual can reach their full potential with dignity and equity within a flourishing and sustainable environment. This approach envisions a world where basic human needs are fulfilled, empowering people to thrive in a just and balanced society.
2. Planet: There will be a determined effort to protect the planet from degradation through sustainable consumption and production, sustainably managing its natural resources and taking urgent action on climate change so that it can support the needs of the present and future generations.
3. Prosperity: All human beings must enjoy prosperous and fulfilling lives in an environment where economic, social and technological progress occurs in harmony with nature.
4. Peace: The agenda will foster peaceful, just and inclusive societies free from fear and violence. There can be no sustainable development without peace or sustainable development.
5. Partnership: The means required to implement this agenda through a revitalised global partnership for sustainable development, based on the

> spirit of strengthened global solidarity, focused in particular on the needs of the poorest and most vulnerable and will be ensured with the participation of all countries, all stakeholders and all people.

The interlinkages and the SDGs' integrated nature are crucial in ensuring that the purpose of the new agenda is realised. If we realise our ambitions across the full extent of the agenda, the lives of all will be profoundly improved and our world will be transformed for the better.

Shared principles and commitments

The SDG Agenda is guided by the purposes and principles of the Charter of the United Nations, including full respect for international law. It is grounded in the Universal Declaration of Human Rights, International Human Rights Treaties, the Millennium Declaration, the 2005 World Summit Outcome Document and the Declaration on the Right to Development. All the other declarations, such as the United Conference on Sustainable Development (Rio+20) and the follow-ups, will be honoured on this agenda. All the goals and commitments in these major conferences and summits are relevant and interrelated. The agenda recognises this, and a new integrated approach has been implemented to solve the commonly accepted issues.

The New Agenda

The New Agenda, compromising seventeen SDGs and 169 associated targets, integrated and inseparable, has brought together countries like never before. The agenda sets a path towards sustainable development of 'win-win' cooperation, which can benefit all member countries worldwide. Gender equality and empowerment of women and girls will contribute significantly to achieving the goals and targets. One half of the population cannot remain deprived of overall development.

The new goals and targets took effect on 1 January 2016 to guide the decisions over the next fifteen years. All countries will work to implement the agenda within their own countries and at the regional and global levels, considering different national realities, capacities and levels of development and respecting national policies and priorities. The United Nations will respect national policy space for sustained, inclusive and sustainable economic growth, particularly for developing states, while remaining consistent with relevant international rules and commitments.

Figure 16 The new agenda. United Nations Foundation, <http://www.unfoundation.org> [accessed on 21 September 2024]

Means of implementation

The scale and ambition of the New Agenda require a revitalised global partnership to ensure its importance. The United Nations is fully committed to making the partnership work in a spirit of international solidarity, particularly with the poorest and most vulnerable people. It will facilitate an intensive global engagement that supports the importance of implementing all the goals and targets and bringing together the governments, the private sector, the civil society, the United Nations and all other actors.

Each country has primary responsibility for its own economic and social development. The means to implement these policies would be:

- Mobilization of financial resources
- Capacity-building
- Transfer of environmentally sound technology to developing countries on favourable terms, including concessional and preferential terms
- Public finance in providing essential services and public goods and in catalysing other sources of financial resources supplemented by Overseas Development Assistance
- Enactment of legislative and adoption of budgets and their role in ensuring accountability

Follow-up and review

To support accountability to the citizens, the United Nations will provide for follow-up and review at various levels, as set out in this agenda and the Addis Ababa Action Agenda.[13] The High Level Political Forum and the Economic and Social Council (ECOSOC) will be central in overseeing follow-up and review at the global level. The work of the Statistical Commission pertaining to the 2030 Agenda for Sustainable Development via a resolution was adopted by the General Assembly on 6 July 2017. The indicators to measure the performance and progress of SDGs will be discussed after a careful study of the SDGs.

Universality and the 2030 agenda for sustainable development

Universality is a defining and innovative feature of the 2030 Agenda for Sustainable Development in its universal nature and scope. It commits all countries to contribute towards a comprehensive effort for global sustainability in all its dimension – social, economic and environmental – while ensuring equity, peace and security.

13 United Nations General Assembly, 'Transforming Our World: The 2030 Agenda for Sustainable Development', Resolution A/RES/70/1, adopted 25 September 2015, <https://www.un.org/sustainabledevelopment/development-agenda> [accessed 25 September 2024].

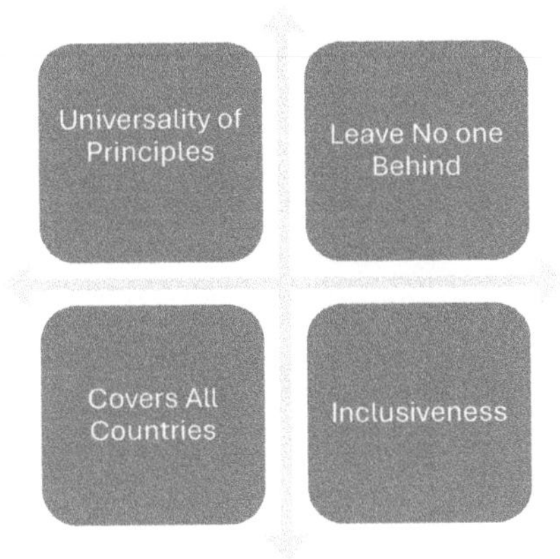

Figure 17 Principle of universality

Principles of universality

Universality is based on four basic principles on how it should be applied in practice. They are shown in Figure 17.

1. Principles: The 2030 Agenda rests on a set of universal principles, values and standards that are applicable in all countries in all contexts and circumstances, at all times. Hence, principles are universal.
2. Leave no one behind: All SDGs are meant for all people, and the call 'leave no one behind' inspired its formulation, constituting a central pledge by member states. This can be understood as a principle of universality of reach.
3. Covers all countries: The 2030 Agenda is not specific to developing countries only. It covers all countries, whether rich or poor, encompassing the set of interlinked trajectories of social, economic and environmental evolution.
4. Inclusiveness: All countries are interconnected at all levels between goals, between countries and between global, regional and national levels. This becomes an inclusive and collective effort for sustainable development. This underscores the idea of the 2030 Agenda as a global partnership for sustainable development.

As a global initiative of universal application, the progress in implementing the 2030 Agenda will need to be monitored in all countries, regardless of income per capita and at all levels.

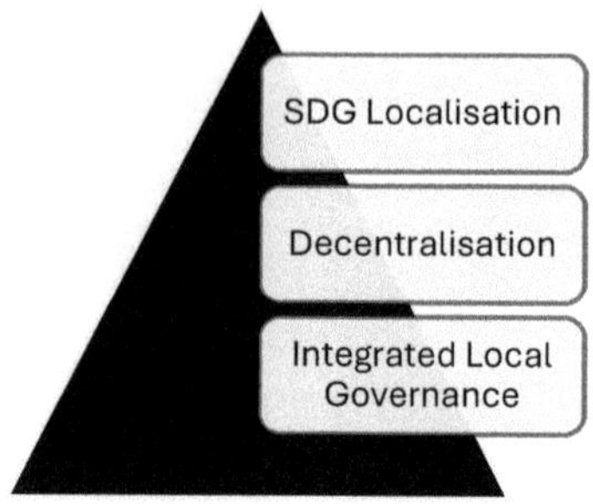

Figure 18 Capacity-building

Capacity building to strengthen sustainable development

Capacity building is essential for strengthening sustainable development by empowering individuals, organizations, and communities with the skills, knowledge, and resources needed to address environmental, social, and economic challenges. It fosters innovation, improves governance, and enhances local resilience, enabling effective implementation of sustainable practices and long-term, inclusive growth. The following are how national and local capacities can be strengthened for sustainable development (Figure 18).

1. SDG localization: The localization of the 2030 Agenda is not the implementation of a global or national agenda at the local level but instead building adequate conditions at the local level to achieve global goals.[14] The focus is on SDG localization, but the most important issue is to determine if the cities and other human settlements are ready with respect to autonomy, capacity, and resources to implement the SDGs effectively in the first place. Local governments do not have the money to take action on SDGs due to the following reasons:
   - Limited political and fiscal power
   - Lack of access to development finance
   - Low levels of institutional capacity
   - The absence of robust multi-level government cooperation and integration
   - The inability to attract or be part of multi-stakeholder solid partnerships

14 United Cities and Local Governments, 'Local Governance for Sustainability', <https://www.uclg.org> [accessed 25 September 2024].

Sustainable governance can be implemented only if these primary constraints are removed.

2. SDGs and decentralised governance: SDG localization opens the window for decentralised governance. Local governments now do public delivery of services in most countries. It leads to more inclusive and representative decision-making. Sustainable development depends on the 'effective decentralization of responsibilities, policy management, decision-making authority and sufficient resources, including revenue collection authority to local authorities, closest to and most representative of, their constituencies'.[15] Even though it seems logical that decentralised governance leads to sustainable development on paper, reality shows that there are uneven development gains with increasing spatial inequalities and exclusion within territories. This wide gap appears due to the lack of capacity and institutional framework of local governments. They need to be upgraded to embark on SDG implementation.

   Steps to be taken by local governments: The local governments must develop their capacities to deal with SDG implementation in the following ways:

   - Information: The local government must ensure that its departments and agencies are well informed with the latest reports and data so that they are aware to put up requisition with respect to more extraordinary legislative, fiscal or executive powers within specific sectors or in operational processes.
   - Performance assessment: They must strive to improve fiscal and expenditure management and performance on essential governance and sector-based indicators.
   - Transparency: All departments must put information on a user-friendly website with free access. Each agency must have a citizens charter and grievance redressal system online for transparency in its processes and operations.
   - Partnership: The local governments must work with the higher levels of government and local communities to have effective accountability.
   - Cooperation: The local governments must cooperate with the neighbouring local governments for mutual benefits and improved development outcomes.
   - Citizen engagement and CS: The local governments must hold meetings, seminars and workshops to develop beneficial networks with civil

15 Ibid.

service organizations and bring more citizens into the fold of sustainable development.

3. Integrated local governance: An integrated governance and policy strategy is essential to implement SDGs successfully. The SDGs have been formed on an international platform, and the UN member countries have pledged to it. Still, they must ultimately be implemented locally to meet its social, economic, and environmental targets. 'Integration of governments' means coordinating working arrangements where multiple departments or public sector organizations are involved in providing a public service or programme.[16]

Elements of an integrated approach: An application of an integrated approach has the following elements:

- Transforming traditional sector-based governance structures
- Promoting coordination between public sector organizations/instruments
- Bringing all development targets and priorities across different levels of governments
- Encouraging mutually beneficial decision-making and reducing trade-offs
- Working out multi-sector planning through joint action

Organizations in SDG cities have highlighted this approach guide. The problem today is that all these governments and agencies work independently, leading to disruptive results and failing SDG implementation.

Conclusion

To conclude the discussion on the transformation of development into sustainable development, we must understand the role of public administration in sustainable development. Development administration is an offshoot of public administration that aims to achieve goal-oriented targets with focused developmental activities and the introduction of SDGs. The 193 member countries of the UN must accept the changed profile of public administration and the changing profile of development administration. Public administration features more prominently in the SDGs than in the MDGs. Having a proper public administration is now a stand-alone development objective. The public governance

16 Christopher Pollitt and Geert Bouckaert, *Public Management Reform: A Comparative Analysis – New Public Management, Governance, and the Neo-Weberian State*, 3rd edn (Oxford: Oxford University Press, 2011).

systems are now explicitly mandated to contribute to realising SDGs through policy instruments.[10] The ultimate attainment of SDGs depends on the public governance systems. SDG Agenda 2030 emphasises three fundamental roles of public administration in SDGs.

First, public administration has the role of a stand-alone development objective.[17] Development is much broader than just providing basic needs to people. It has broadened its scope to include security, citizenship, and the fundamental rights of human beings, as is evident from SDG 16. Most of the targets of sustainable development directly relate to the policy instruments of public administration, such as women's participation in political decision-making and PPPs. Second, policy instruments of public administration need to be re-evaluated. The SDGs are designed explicitly with respect to the policy instruments that public governance should mobilise to transform development administration into sustainable development, representing a profound shift in governance paradigms, policy frameworks, and societal aspirations. From its origins rooted in traditional economic growth and social progress models, development administration has evolved into sustainable development governance to embrace the principles of sustainability, equity, and resilience.

Public governance is the key to delivering these services to the citizens. For example, the government can provide the required number of buses to a state, but if these do not run on time, the governance is to be blamed. The policies, strategies, or measures for SDGs need to be formulated by a responsive public administration system that focuses on gender equality, water and sanitation, job creation, and so on. Public administration needs to cater to all SDGs by mobilizing resources through sound mechanisms which have a legislative base. It must ensure that the developing countries have a proper representation in the international forums to harness global support. Third, the SDGs are embedded and framed intrinsically in the context of public governance. They comprise the interplay between the public sector, the private sector and the non-profit sector to achieve the objectives of policies.

This transformation has been driven by recognising the interconnectedness between environmental health, social well-being, and economic prosperity. It reflects a growing awareness of the finite nature of natural resources, the urgency of addressing global challenges such as climate change and biodiversity loss, and the imperative to build inclusive and resilient societies that leave no one behind. Key factors contributing to this transformation include advances in environmental science, shifts in public attitudes towards sustainability, and the

---

17 Francis Fukuyama, 'What is Governance?', *Governance*, 26.3 (2013), 347–368.

emergence of international agreements and frameworks such as the Sustainable Development Goals (SDGs). These milestones have catalysed efforts to integrate sustainability principles into policy-making processes, institutional structures, and development agendas at local, national, and global levels.

The journey from development administration to sustainable development has not been without challenges. It has required navigating complex trade-offs between economic growth and environmental protection, balancing short-term priorities with long-term sustainability goals, and overcoming institutional inertia and vested interests resistant to change. However, despite these challenges, the transition towards sustainable development offers immense opportunities for innovation, collaboration, and transformative change. It invites us to rethink traditional notions of progress and prosperity, embrace new models of governance and decision-making that prioritise the well-being of both present and future generations, and to cultivate a deeper sense of stewardship and responsibility towards our planet and its inhabitants.

As we continue on this journey, it is essential to remain committed to the principles of sustainability, equity and inclusivity. It requires fostering partnerships, sharing knowledge, and empowering communities to actively participate in shaping their own sustainable futures. We can realise the promise of sustainable development and build a more just, resilient, and flourishing world for all by working together with vision, determination, and collective action.

Models of Sustainable Development in Action

From Theory to Practice

Introduction

Sustainable development governance is central to achieving long-term sustainability goals at both national and global levels. Governance in this context refers to the systems, institutions, policies and processes through which decisions about sustainable development are made, implemented and monitored. The effectiveness of these governance structures significantly impacts a country's ability to balance economic growth, social equity and environmental protection, ultimately determining the success of its sustainable development efforts. Integration of the economic, social and environmental pillars of sustainability into a coherent and cohesive policy framework is at the heart of sustainable development governance. This integration is critical because the three pillars are inherently interconnected, and progress in one area often depends on advances in the others. Effective governance ensures that policies across these domains are not developed in isolation but are instead coordinated to promote synergies and avoid trade-offs. For instance, economic policies must be designed to encourage growth that is both inclusive and environmentally sustainable, while social policies should aim to reduce inequality in ways that do not deplete natural resources.[1]

One of the key challenges in sustainable development governance is the need for robust institutional frameworks to support the implementation of sustainability policies. These frameworks must coordinate actions across different sectors and levels of government and engage with non-state actors such as businesses, civil society organizations and local communities. Effective institutions are characterised by transparency, accountability and the ability to enforce regulations and standards. They must also be flexible and adaptive, able to respond to changing circumstances and emerging challenges, such as those posed by

1 Frank Biermann, Norichika Kanie, and Rakhyun E Kim, 'Global governance by goal-setting: the novel approach of the UN Sustainable Development Goals', *Current Opinion in Environmental Sustainability* 26–27 (2017), 26–31.

climate change or economic disruptions.[2] The role of leadership and political will is crucial in sustainable development governance. Leaders prioritizing sustainability are more likely to drive the necessary reforms and investments to achieve sustainable outcomes. Political will is often demonstrated through enacting comprehensive sustainability legislation, allocating adequate resources for implementation and establishing clear goals and timelines. Leaders must foster a culture of sustainability within government institutions and society at large, promoting values such as stewardship, responsibility and intergenerational equity.[3]

Public participation and stakeholder engagement are vital components of effective, sustainable development governance. Involving diverse stakeholders, particularly those who are marginalised and vulnerable, guarantees that sustainability initiatives are inclusive and address the needs and goals of all societal groups. Public participation enhances the legitimacy of governance processes and increases the likelihood of successful policy implementation by building broad-based support and ownership. Mechanisms for stakeholder engagement can include public consultations, participatory planning processes and the establishment of multi-stakeholder platforms that bring together government, private sector and civil society representatives.[4] Integrating sustainability into national planning and budgeting processes is essential for good sustainable development governance. This involves aligning national development plans with sustainability goals, such as the United Nations' Sustainable Development Goals (SDGs), and ensuring that budgets reflect the prioritization of sustainable initiatives. It also requires the development of metrics and indicators to monitor progress and assess the impact of policies on sustainability outcomes. Countries can ensure that sustainable development is not treated as an add-on but is central to their development strategies by embedding sustainability into the core functions of government.[5]

Global cooperation and international partnerships are also essential in sustainable development governance. Sustainability development challenges, such as

---

2 Felix Dodds, David Donoghue, and Jimena Leiva Roesch, *Negotiating The Sustainable Development Goals: A Transformational Agenda For An Insecure World* (London: Routledge, 2016).

3 Jon Meadowcroft, 'Who is in charge here? Governance for sustainable development in a complex world', *Journal of Environmental Policy & Planning* 9.3-4 (2007), 299–314.

4 United Nations Development Programme, *Governance for Sustainable Development: Integrating Governance in the Post-2015 Development Framework* (New York: United Nations Development Programme, 2019).

5 Oran R Young, 'Governing complex systems: social capital for the Anthropocene', *Global Environmental Politics* 17.2 (2017), 40–48.

climate change, biodiversity loss and pollution are transboundary in nature and require collective action at the international level. Countries need to collaborate through international agreements, share best practices and support each other in building the capacity to implement sustainable development policies. Global governance structures, such as the United Nations and its various agencies, play a key role in facilitating these partnerships and providing platforms for dialogue and cooperation. Thus, through robust governance frameworks, countries can successfully navigate sustainable development challenges and achieve long-term sustainability goals that benefit current and future generations.

In the global village where we live at present, it is in everyone's interest that any society ideally meets the conditions mentioned above. However, some societies have difficulties achieving those conditions despite actively seeking and working towards these goals. On the other hand, some societies have the capability and means to assist others because of their experience, working knowledge, and available resources. Therefore, a proactive effort to achieve such conditions should become a prime vehicle for global cooperation. The Good Governance in Sustainable Development (GGSD) Program aims to assist societies in developing an effective government within a democratic system and implement sustainable development principles through global partnership.[6]

Integrating governance into the Sustainable Development Goals (SDG) agenda

Integrating governance into the Sustainable Development Goals (SDG) agenda is essential for ensuring that the goals are effectively implemented, monitored, and achieved. In this context, governance refers to the processes, institutions, and systems that guide decision-making, accountability and participation at various local, national and global levels. Several approaches have been developed to integrate governance into the SDG agenda. What follows is a closer look at each approach.

1. Incorporating governance as a standalone goal: The inclusion of SDG 16, which focuses on promoting peaceful and inclusive societies, providing access to justice for all and building effective, accountable and inclusive institutions at all levels, is a crucial approach to integrating governance into the SDG agenda. SDG 16 explicitly recognises the importance of

6 United Nations, *Transforming Our World: The 2030 Agenda for Sustainable Development* (New York: United Nations, 2015), <https://sdgs.un.org/2030agenda>, accessed 29 August 2024.

governance in achieving sustainable development, highlighting the need for transparent, accountable and inclusive institutions.[7]

2. Mainstreaming governance across all SDGs: Governance is not only confined to SDG 16 but is also a cross-cutting theme that affects the achievement of all other SDGs. Effective governance is necessary to create an enabling environment for implementing each goal. For example, governance is critical in achieving SDG 1 (No Poverty) by ensuring that resources are distributed equitably and that social protection systems are in place.[8] Similarly, achieving SDG 13 (Climate Action) requires governance frameworks that promote international cooperation, enforce environmental regulations and support adaptive management practices.[9]
3. Strengthening multi-level governance: Sustainable development challenges often transcend national borders and require coordinated action at multiple levels: local, national, regional and global. Strengthening multi-level governance involves creating mechanisms for cooperation and coordination among different levels of government and between governments and other stakeholders. This approach ensures that global goals are translated into local actions and that local needs and priorities are reflected in national and global policies.[10]
4. Promoting inclusive and participatory governance:: Inclusive and participatory governance is essential for ensuring that the voices of all stakeholders, particularly marginalised and vulnerable groups, are heard in the decision-making process. This approach involves creating platforms and mechanisms for meaningful participation, such as multi-stakeholder partnerships, public consultations, and community-based monitoring. Governance frameworks can ensure that the SDG agenda is responsive to the needs and aspirations of all segments of society by promoting inclusivity and participation.[11]

---

7 Jeffrey D Sachs, 'From Millennium Development Goals to Sustainable Development Goals', *The Lancet* 379/9832 (2012), 2206–2211.

8 United Nations Development Programme, *Localising the Sdgs: Supporting Localisation through Multi-Level Governance* (New York: United Nations Development Programme, 2019).

9 John Gaventa, 'Finding the spaces for change: a power analysis', *IDS Bulletin* 37.6 (2006), 23–33.

10 Organisation for Economic Co-operation and Development, *Better Policies for Sustainable Development 2016: A New Framework for Policy Coherence* (Paris: OECD Publishing, 2016).

11 World Bank, *Strengthening Institutional Capacity for Sustainable Development* (Washington, DC: World Bank, 2019).

5. Leveraging digital governance tools: Digital technologies and e-governance tools have emerged as an essential approach to enhancing governance for sustainable development. Digital governance tools like open data platforms, digital monitoring systems and online public consultations can improve transparency, accountability, and citizen engagement. These tools can help track progress toward the SDGs, facilitate data-driven decision-making and enable real-time stakeholder feedback.
6. Integrating governance indicators in SDG monitoring: Integrating governance indicators into the monitoring and evaluation frameworks for the SDGs is crucial for assessing the effectiveness of governance systems in achieving the goals. Governance indicators can measure various aspects of governance, such as the rule of law, corruption, institutional capacity, and civic participation.[12] Governments and international organisations can identify governance gaps and plan targeted interventions to address them by tracking these indicators.
7. Promoting policy alignment: Achieving the SDGs requires coherent policies that align with sustainable development objectives across sectors. Policy coherence ensures that trade, finance, agriculture, and energy policies are consistent with sustainable development goals and do not undermine progress in other areas.[13] This approach necessitates robust governance frameworks facilitating intersectoral coordination and integration, enabling policymakers to address trade-offs and synergies among different SDGs.
8. Building institutional capacity: Building the capacity of institutions to implement, monitor and enforce sustainable development policies is a critical approach to integrating governance into the SDG agenda.[14] This includes strengthening the capacity of public institutions to deliver services effectively, enhancing the rule of law, and improving the ability of institutions to engage with stakeholders and respond to emerging challenges.

Integrating governance into the SDG agenda requires a multi-dimensional approach that addresses governance at all levels and across all goals. By incorporating governance as a standalone goal, mainstreaming it across the SDGs,

12 Kate Raworth, *Doughnut Economics: Seven Ways to Think Like a 21st-century Economist* (London: Random House Business, 2017).

13 Herman E. Daly and Joshua Farley, *Ecological Economics: Principles and Applications*, 2nd edn (Washington, DC: Island Press, 2011).

14 Paul James, Liam Magee, Andy Scerri and Manfred B Steger, *Urban Sustainability in Theory and Practice: Circles of Sustainability* (London: Routledge, 2015).

strengthening multi-level and participatory governance, leveraging digital tools, integrating governance indicators, promoting policy coherence, and building institutional capacity, the global community can ensure that governance supports the achievement of sustainable development.

Models of sustainable development: Integrating economic, social, and environmental goals

Sustainable development is a concept that has evolved to address the growing challenges of meeting present needs without compromising the ability of future generations to meet their own. Central to this concept are various models that provide frameworks for balancing economic, social and environmental goals. These models guide policymakers, businesses and communities in achieving sustainability. This essay explores some critical models of sustainable development, examining their origins, principles, applications, and how they contribute to a sustainable future.

The 'Three Pillars Model'

The 'Three Pillars Model', also known as the 'Triple Bottom Line', is perhaps the most recognised and widely used model of sustainable development. This model emphasises balancing the three interconnected pillars: economic growth, social inclusion and environmental protection. The idea is that for development to be sustainable, it must address all three aspects simultaneously, ensuring that progress in one area does not come at the expense of another.

Key Features

The features of the Three Pillars Model include the three dimensions of sustainable development. They are briefly described here and given in Figure 19.

1. Economic sustainability: The economic pillar focuses on generating long-term prosperity through inclusive economic growth. It emphasises that

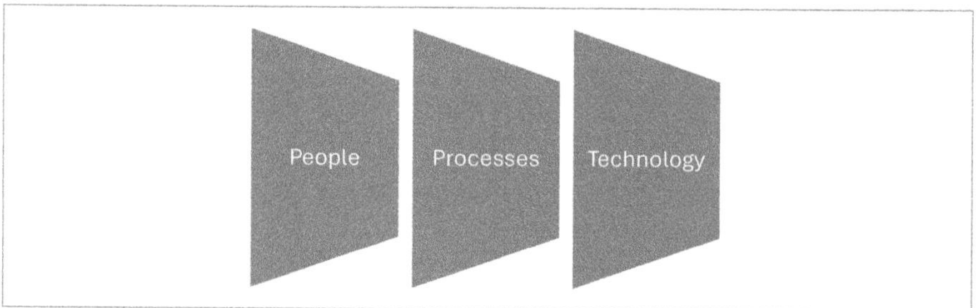

Figure 19 Three Pillars Model

economic development should be about increasing GDP and ensuring that the benefits of growth are widely shared. This includes creating jobs, reducing poverty and fostering innovation. Maintaining economic sustainability requires that economic activities avoid exhausting natural resources or causing environmental damage that could jeopardise future economic opportunities.

2. Social sustainability: Social sustainability is concerned with improving the quality of life for all members of society. It involves addressing issues such as inequality, poverty and social exclusion. Social sustainability also includes promoting education, healthcare and social justice. The goal is to create a society that is equitable and inclusive, where everyone has the opportunity to thrive.
3. Environmental sustainability: The environmental pillar protects and conserves natural resources and ecosystems. It involves reducing pollution, mitigating climate change and ensuring the sustainable use of resources such as water, land, and biodiversity. Environmental sustainability also includes the concept of intergenerational equity, ensuring that future generations have access to the resources they need to live well.

However, the model has been criticised for being overly simplistic, as it does not fully capture the complex interrelationships between economic, social, and environmental factors. This model is also called the 'Three Circles Model' and is based on human society but does not consider the 'human quality of life'. However, some concerns have been raised by researchers against this Model. They are:

- Separation of human and environmental development: The premise that the three pillars are independent constructs is countered by the argument that human resources are biological entities and, therefore, are also environmental and cannot be independent. Since majority of resources come from nature in the form of ecosystems, it is impossible to separate human development from environmental development. For example, if we destroy ecosystems for agriculture, we potentially destroy access to natural resources like food, wood and medicines.
- Time dimension: The Three Pillar Model does not incorporate a time dimension.
- Content: The content of the three pillars individually lacks consensus.

The Three Pillars Model has been adopted by governments, businesses and organizations worldwide as a guiding framework for sustainable development. It

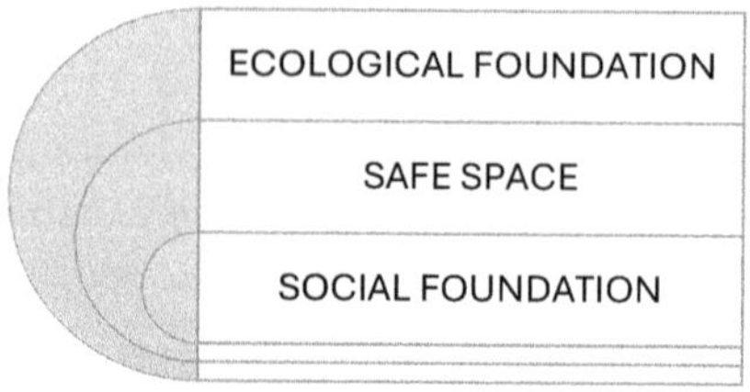

Figure 20 Doughnut Economics Model

is often used in policymaking, corporate social responsibility (CSR) initiatives, and community development projects.

Doughnut Economics Model

Doughnut Economics Model, proposed by economist Kate Raworth, offers a fresh perspective on sustainable development by combining the concept of planetary boundaries with social foundations. This model visualises a doughnut-shaped space where human needs are met without exceeding the Earth's ecological limits. The model challenges the traditional focus on economic growth and instead advocates for a safe and just space for humanity.[15] Doughnut Economics Model can be understood better through Figure 20.

Key Features: Three main features of Doughnut Economics help understand the model. They are:

- Inner Circle: The inner circle of the doughnut represents the minimum living standards required to ensure social equity. It denotes access to essential resources such as food, water, healthcare and education. The inner circle is aligned with the concept of social foundations necessary for a decent quality of life.
- Outer Circle: The outer circle represents the ecological ceiling that must not be breached to avoid environmental degradation. This includes the nine planetary boundaries identified by scientists, such as climate change, biodiversity loss and ocean acidification. Breaching these boundaries could lead to catastrophic environmental consequences that threaten human survival.

Safe and Just Space: The area between the inner and outer circles is the 'safe and just space' where sustainable development occurs. This space represents a balance between meeting human needs and respecting the Earth's ecological

15 Karl-Henrik Robert, *The Natural Step Story: Seeding a Quiet Revolution* (Gabriola Island, BC: New Society Publishers, 2002).

limits. Doughnut Economics emphasises the need to rethink economic systems and prioritise well-being over continuous economic growth.

Doughnut Economics has gained traction among cities and organizations seeking to implement sustainable practices. For example, Amsterdam has adopted the model to guide its sustainability policies, aiming to create a circular economy that minimises waste and maximises resource efficiency. The model's social and environmental equity emphasis has resonated with many sustainable development advocates. However, critics argue that the model may be challenging to implement on a global scale due to different countries' varying capacities to meet social and ecological goals.

Ecological Economics

Ecological Economics is a model that emphasises the interdependence of human economies and natural ecosystems. Unlike traditional economics, which often treats the environment as an external factor, Ecological Economics recognises that the economy is embedded within the environment and must operate within its carrying capacity. This model advocates for a steady-state economy that prioritises sustainability over continuous growth.

Features of Ecological Economics Model

This model has three features, as shown below.

- Steady-state economy: The concept of a steady-state economy is central to Ecological Economics. It requires achieving a stable level of economic activity that aligns with the Earth's ecological limits. The steady-state economy focuses on maintaining a balance between economic production and the regenerative capacity of the environment.
- Natural capital: Ecological Economics recognises the value of natural capital, which includes ecosystem services, biodiversity, and other natural resources. The model advocates for preserving and conserving natural capital as essential to human well-being. It also emphasises the need to account for environmental costs in economic decision-making.
- Carrying capacity: The model highlights the importance of understanding and respecting the Earth's carrying capacity, which refers to the maximum population size and level of resource consumption that the environment can sustainably support. Ecological Economics calls for policies and practices that ensure human activities remain within these limits.

Ecological Economics has influenced environmental policies and sustainable development initiatives worldwide. It has informed approaches such as the Green

Economy and Circular Economy, which seek to decouple economic growth from environmental degradation. The model's focus on the intrinsic value of nature has also contributed to the development of concepts such as ecosystem services and natural capital accounting. However, ecological economics faces challenges in gaining widespread acceptance, particularly in economies prioritizing growth over sustainability.

'Circles of Sustainability' model

The 'Circles of Sustainability' model, developed by the United Nations, offers a comprehensive approach to assessing sustainability across four dimensions: economic, ecological, political, and cultural. This model emphasises the need for a holistic understanding of sustainability, recognizing that economic and environmental factors are intertwined with social and cultural contexts.

Features of 'Circles of Sustainability' model

The essential features of this model are briefly described here.

- Economic: The economic dimension of the Circles of Sustainability focuses on production, exchange and consumption practices that are sustainable in the long term. It emphasises the importance of creating resilient economies and adapting to changing environmental and social conditions.
- Ecological: The ecological dimension examines the sustainability of interactions between human activities and the environment. It necessitates the sustainable use of natural resources, conservation of biodiversity, and mitigation of environmental impacts such as pollution and climate change.
- Political: The political dimension of the model addresses governance, power, and decision-making processes that support sustainability. It highlights the importance of inclusive and transparent governance systems that enable effective environmental and social policies.
- Cultural: The cultural dimension emphasises the role of cultural practices, values and diversity in achieving sustainable development. It recognises that cultural identity and heritage are integral to the social fabric and contribute to the resilience of communities.

The 'Circles of Sustainability' model has been used in urban planning, community development and sustainability assessments. It has been particularly influential in developing sustainable cities, where integrating economic, ecological, political and cultural factors is essential for creating liveable urban environments. The model's holistic approach makes it a valuable tool for policymakers and planners. However, its complexity can be a barrier to implementation, as it requires a deep understanding of the interconnectedness of various sustainability dimensions.

The 'Natural Step Framework Model'

The Natural Step Framework, developed by Swedish scientist Karl-Henrik Robert, is a science-based model for achieving sustainability. The framework is built on four system conditions defining the boundaries for sustainable human activities. The model advocates for a strategic approach to sustainability through a process called backcasting.[18]

Features

The Natural Step Framework Model has the following vital features for its implementation.

- Four system conditions: The four system conditions of the Natural Step Framework are designed to prevent environmental degradation and ensure sustainability. These conditions include reducing dependence on fossil fuels and other non-renewable resources, minimising the use of toxic substances, reducing waste and emissions and meeting human needs reasonably and efficiently.
- Backcasting: Backcasting is a crucial methodology of the Natural Step Framework. It involves envisioning a sustainable future and then planning the steps needed to achieve that future from the present. This approach encourages long-term thinking and strategic planning, helping organizations and communities to align their actions with sustainability goals.
- Science-based approach: The Natural Step Framework is grounded in scientific principles, providing a rigorous basis for sustainability planning. The model emphasises the importance of understanding the biophysical limits of the planet and the need for human activities to operate within these limits.

Businesses, governments and non-profit organizations worldwide have adopted the Natural Step Framework. It has been used to guide sustainability initiatives in manufacturing, energy and urban development sectors. The framework's emphasis on scientific rigour and strategic planning makes it a valuable tool for organizations seeking long-term sustainability. However, the model's reliance on backcasting may be challenging for organizations accustomed to short-term planning and reactive decision-making.

'Sustainable Livelihoods Framework Model'

The Sustainable Livelihoods Framework is a model often used in development contexts, particularly in rural areas and low-income communities. The model focuses on enhancing the resilience of communities by strengthening their assets and reducing vulnerability. It emphasises the importance of a holistic approach.

The diverse models of sustainable development discussed above offer valuable frameworks for addressing the complex challenges of integrating economic, social and environmental goals. Each model, the widely recognised Three Pillars Model, the innovative Doughnut Economics or the holistic Circles of Sustainability, provides unique insights and strategies for achieving sustainability. While no single model may be universally applicable, it offers a comprehensive toolkit for governments, businesses, and communities striving to create a sustainable future. The choice of model often depends on the specific context and objectives. Still, the common thread across all models is the commitment to ensuring that development today does not compromise the well-being of future generations. As sustainability continues to rise on the global agenda, these models will play an increasingly critical role in guiding policies and practices that promote a balanced, resilient, and equitable world.

Evaluating the optimal model for sustainable development

When it comes to adopting a model for sustainable development, each country faces unique challenges and opportunities that influence its decision-making process. Choosing the best model depends on factors like the economy, society, environment, politics and culture. It is not a one-size-fits-all decision. This evaluation process is critical to ensuring that the chosen model aligns with the country's specific context and can effectively address its sustainability goals.

The major points to be considered in evaluating the optimal model for sustainable development are listed in Figure 21.

1. Economic considerations: First and foremost, economic considerations play a pivotal role in determining which model a country might adopt. Countries with robust economies may lean towards models that emphasise economic sustainability alongside social and environmental considerations, such as the Three Pillars Model. This model allows for a balanced approach where economic growth is pursued with social inclusion and environmental protection. In contrast, developing nations or those with fragile economies might prioritise models focusing on poverty alleviation and resource conservation, such as the Sustainable Livelihoods Framework. This model offers a pathway to enhance resilience by strengthening local

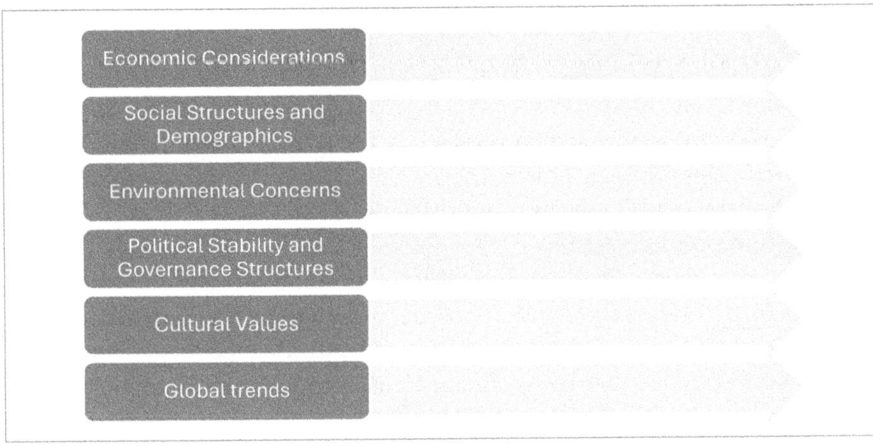

Figure 21 The optimal model for sustainable development

assets and reducing vulnerability, essential in contexts where economic stability is precarious.

2. Social structures and demographics: Social structures and demographics are also significant factors in this decision-making process. Countries with high levels of inequality or diverse populations may find models like Doughnut Economics particularly appealing. This model incorporates planetary boundaries and emphasises social equity, ensuring that all citizens have access to a minimum standard of living while avoiding environmental degradation. For nations facing social unrest or deep-seated inequality, a model that explicitly addresses social foundations and justice will likely be more relevant and effective.
3. Environmental concerns: Environmental concerns are another crucial consideration. Countries highly vulnerable to climate change, such as small island nations or those with large coastal populations, may prioritise models emphasising ecological boundaries and environmental sustainability. Ecological Economics, which places the economy within the context of environmental limits, could be a preferred choice for these nations. The model's focus on maintaining a steady-state economy within the Earth's carrying capacity resonates with the urgent need to protect natural resources and ecosystems for these countries' survival.
4. Political stability and governance: Political stability and governance structures also heavily influence the selection of a sustainable development model. In countries with strong, transparent, and inclusive governance, models like the Circles of Sustainability, which incorporate political dimensions, may be more suitable. This model emphasises the importance

of governance, power, and decision-making processes in achieving sustainability. On the other hand, in nations where governance is weak or authoritarian, models that can be implemented with minimal reliance on complex governance structures, such as The Natural Step Framework, may be more appropriate. This framework's focus on science-based principles and strategic planning through backcasting can be adapted to varying levels of political stability.

5. Cultural values and societal priorities: Cultural values and societal priorities further shape the decision-making process. Nations with deep-rooted cultural traditions or those that highly value cultural identity may gravitate towards models that recognise and integrate cultural dimensions, such as the Circles of Sustainability. This model's inclusion of cultural factors ensures that sustainable development strategies are culturally sensitive and resonate with the population's values and beliefs. In contrast, countries with less prominent cultural factors might prioritise models that focus more on economic and environmental sustainability.
6. Global trends: Finally, the decision-making process is also influenced by global trends, international partnerships, and the availability of external support. Countries that are part of global agreements or that receive international aid may adopt models that align with global sustainability frameworks, such as the Sustainable Development Goals (SDGs). These global influences can guide national decisions towards internationally recognised and supported models, ensuring that national strategies are consistent with global sustainability efforts.

In summary, selecting a sustainable development model for any country is complex and multifaceted. It requires a careful evaluation of economic, social, environmental, political and cultural factors and consideration of global trends and partnerships. The chosen model must be adaptable to the country's specific needs and conditions to promote sustainability and improve the well-being of its population effectively. When a country decides which model of sustainable development to adopt, it typically considers various factors, including its unique economic, social, environmental, and political contexts.

Country-specific model

The process of the selection of a model for a specific country is often complex and involves multiple stakeholders. Here's how a specific country might go about determining which model is the best fit:

1. Assessment of national context: The assessment of a national context is crucial in selecting an appropriate sustainable development model for a specific country. This process involves analysing the country's unique socioeconomic conditions, environmental challenges, governance structures and cultural values. Policymakers can tailor sustainable development strategies that align with the country's specific needs and priorities by understanding these factors. Moreover, this assessment helps identify existing resources, potential barriers, and opportunities for implementing sustainable practices. A thorough national context assessment ensures that the chosen model addresses immediate development goals and fosters long-term sustainability, resilience, and equitable growth. The conditions for the evaluation of the national context are given below.
   - Economic conditions: Countries assess their economic priorities, such as growth, employment, industrialization, and infrastructure needs. A country heavily reliant on natural resources might prioritise models that focus on sustainable resource management.
   - Social factors: Issues like poverty, inequality, health, education and social cohesion play a critical role. A country with significant social disparities might lean towards models emphasising social inclusion and equity.
   - Environmental challenges: Countries consider their environmental context, including biodiversity, climate risks, pollution levels and natural resource availability. Nations facing severe environmental degradation might prioritise models like the Planetary Boundaries Model or Ecological Economics.
   - Political and Institutional Frameworks: The strength and nature of governance, legal systems and political will influence which models can be effectively implemented. A country with robust institutions might be more inclined to adopt comprehensive models like the Circles of Sustainability.
2. Alignment with international commitments: Selecting a model for sustainable development governance requires careful alignment with international commitments to ensure coherence and effectiveness. Integrating global standards, principles and frameworks, such as the United Nations Sustainable Development Goals (SDGs), into national strategies is essential. This alignment leverages global expertise, attracts international support and addresses transboundary issues more effectively. It enhances the credibility and legitimacy of national policies while fostering a collaborative approach to global challenges. A well-aligned model can drive more inclusive and sustainable development outcomes tailored to a country's unique context, all while meeting its international obligations. Many countries align their sustainable development strategies with international frameworks like

the United Nations Sustainable Development Goals (SDGs). The SDGs are based on the Three Pillars Model, so countries often integrate this model into their national policies to meet their international obligations.

3. Stakeholder consultation: Stakeholder consultation is a critical component in selecting a model for sustainable development governance, ensuring that diverse perspectives and interests are considered. Engaging various stakeholders, including government agencies, private sector representatives, civil society organizations and local communities, helps identify critical issues, need and potential challenges. This inclusive approach fosters greater buy-in and support for the chosen model, enhances its relevance, and improves its implementation. Countries can develop more effective and equitable policies that address the unique needs of different groups, ultimately leading to more sustainable and successful outcomes by incorporating feedback from all relevant parties. The appropriate parties in this perspective are:
   - Government agencies: Various ministries, including the ministries of environment, finance, and social affairs, provide input based on their specific mandates.
   - Civil society and NGOs: Public consultations and input from non-governmental organizations can shape which model is deemed appropriate, ensuring that the needs of various segments of society are considered.
   - Private sector: Businesses and industries often advocate for models considering economic sustainability and might push for frameworks like the Triple Bottom Line.
   - Academia and experts: Input from scholars and researchers can provide evidence-based recommendations on which models might work best in the national context.
4. Pilot projects and case studies: Pilot projects and case studies are crucial for selecting and refining sustainable development governance models. Pilot projects act as experimental platforms to test and validate theoretical models in real-world settings, offering valuable insights into their practicality, effectiveness and potential challenges. Implementing these small-scale projects allows policymakers to assess the feasibility of various approaches, gather empirical data, and make necessary adjustments before broader implementation. Case studies comprehensively analyse how different models have been applied in multiple contexts. They offer lessons learned, best practices and cautionary tales from similar situations in other countries or regions. Analysing case studies helps identify successful strategies, potential pitfalls, and contextual factors influencing the model's effectiveness.

Combining insights from pilot projects and case studies leads to a more informed decision-making process. This approach enables models to adapt to specific national contexts, leveraging proven solutions while avoiding known issues. The result is a higher likelihood of successful implementation and sustainable development outcomes grounded in practical experiences and evidence-based insights. Countries often test different approaches through pilot projects, regional initiatives, or sector-specific strategies. Successes and lessons learned from these pilots can inform the broader adoption of a particular model.

5. Comparative Analysis: Comparative analysis is an essential method in selecting a model for sustainable development governance, involving systematically evaluating different models to determine their suitability for a specific country. This approach examines various models implemented in other countries or regions, comparing their effectiveness, efficiency and adaptability to diverse contexts. Critical factors in this analysis include governance structures, policy frameworks, stakeholder engagement mechanisms and outcomes achieved. Policymakers can identify best practices and successful strategies that have been effective in similar contexts through comparative analysis. It also helps to uncover potential challenges and limitations associated with different models, providing insights into how these issues might be addressed. This method allows for assessing how well various models align with a country's specific needs, resources, and socio-economic conditions.

   This approach ensures that the selected model is theoretically sound but also practical and effective in the country's unique context. Countries can enhance the likelihood of successful implementation and achieve more sustainable development outcomes tailored to their specific requirements by learning from the experiences of others and adapting successful elements to local circumstances. Governments may examine how similar countries or regions have implemented sustainable development models. This can comprise studying case studies, best practices, and outcomes from other countries with comparable economic, social, and environmental conditions.

6. Policy coherence and integration: Policy coherence and integration are essential considerations when selecting a model for sustainable development governance. Ensuring that the chosen model aligns with existing policies and strategies helps create a unified and efficient approach to development. Policy coherence involves harmonising various policies across different sectors to avoid conflicts and duplications, which can lead to more effective and streamlined implementation. Integration focuses on

embedding sustainable development principles into all relevant areas of policy and governance. This requires coordination among various government departments, agencies and stakeholders to align their efforts and resources with the selected model. Effective integration ensures that sustainable development objectives are reflected in economic, social, and environmental policies, enhancing their overall impact.

Assessing policy coherence and integration involves evaluating how well the proposed model complements and strengthens current policies. It also includes identifying potential gaps or overlaps that might hinder progress. A well-integrated model facilitates a more holistic approach to addressing complex development challenges, leveraging synergies between different policy areas. This approach maximises resource utilization and enhances the effectiveness of development initiatives, leading to more sustainable and coordinated outcomes.

An example of policy coherence and integration in selecting a model for sustainable development is the European Union's 'Europe 2020' strategy. This strategy aimed to align economic, social and environmental policies to promote smart, sustainable and inclusive growth across member states. The Europe 2020 strategy integrated various policy areas, including innovation, education, climate change and employment, into a cohesive framework. The strategy ensured that policies across different sectors were mutually supportive and reinforced each other by setting specific targets and priorities in these areas. For instance, efforts to reduce greenhouse gas emissions were aligned with initiatives to promote renewable energy and improve energy efficiency. Similarly, economic policies to stimulate growth were integrated with social policies to address poverty and social exclusion.[16] This approach enhanced policy coherence by avoiding conflicting objectives and ensuring that all policies contributed to the overarching goals of sustainable development. The integration of diverse policy areas allowed for a more coordinated and effective implementation, demonstrating how aligning policies can drive comprehensive and sustainable outcomes. Thus, the chosen model must be integrated into existing national policies and development plans. This ensures that sustainable development is not treated as a separate agenda but is embedded in broader economic and social policies.

16 European Commission, *Europe 2020: A Strategy for Smart, Sustainable and Inclusive Growth, Com (2010) 2020 Final* (Brussels: European Commission, 2010).

7. Flexibility and adaptability: Flexibility and adaptability are crucial attributes when selecting a model for sustainable development governance. These qualities ensure that the chosen model can respond effectively to changing conditions, emerging challenges and evolving priorities. A flexible model allows for adjustments based on new information, unexpected developments, or shifts in socio-economic or environmental contexts. Adaptability involves tailoring the model to address specific local needs while remaining resilient to external pressures and changes. An illustrative example is New Zealand's approach to environmental management. The country implemented the Resource Management Act (RMA) in 1991, which provides a flexible framework for managing natural resources. The RMA allows for adjustments to regulations and practices in response to new scientific data, changing environmental conditions, and community needs. This adaptability has enabled New Zealand to address diverse environmental challenges while aligning with sustainable development goals.[17] Thus, effective flexibility and adaptability in a model contribute to its long-term success and relevance. Ensuring that the model can evolve in response to dynamic conditions enhances its effectiveness and sustainability. Often, countries do not strictly adhere to one model but instead adopt a hybrid approach, tailoring elements from various models to best meet their specific needs. For example, a country might combine the Three Pillars Model with aspects of Doughnut Economics to address both social equity and environmental limits.
8. Monitoring and evaluation: Monitoring and evaluation are integral to selecting and implementing a model for sustainable development governance. These processes involve continuously assessing the performance and impact of the model to ensure it meets its objectives and adapts to changing circumstances. Effective monitoring provides real-time data on progress and helps identify any deviations from expected outcomes. Evaluation, on the other hand, offers a more comprehensive analysis of the model's effectiveness and impact over time, guiding necessary adjustments and improvements. For example, the use of the Results-Based Management (RBM) framework by the United Nations Development Programme (UNDP) emphasises systematic monitoring and evaluation to measure the success of development initiatives. The UNDP tracks progress using performance indicators, evaluates results against set targets, and adjusts strategies based

17 Ministry for the Environment, Resource Management Act 1991 (Wellington: Ministry for the Environment, 2020), <https://www.mfe.govt.nz/rma>, accessed 29 August 2024.

on findings through RBM. This approach ensures that development models are continually refined and aligned with evolving needs and priorities. Hence, monitoring and evaluation enhance the robustness and relevance of a model, fostering evidence-based decision-making and improving overall outcomes. Once a model is selected and implemented, continuous monitoring and evaluation are crucial. This allows for adjustments and adaptations based on what is working or where there are gaps. Countries may revise their approach over time, incorporating lessons learned and new developments.

9. Political will and leadership: Political will and leadership are essential factors in selecting and implementing a model for sustainable development governance. Political will refers to the commitment of government leaders and institutions to prioritise and support sustainable development initiatives. Strong leadership ensures that there is a clear vision, effective coordination and the allocation of necessary resources to achieve the desired outcomes. This commitment is crucial for overcoming obstacles, securing stakeholder buy-in, and driving the model's success. For instance, effective political will and leadership is the commitment of Costa Rica to environmental sustainability. Under the leadership of former President Luis Guillermo Solís, Costa Rica pursued an ambitious goal to achieve carbon neutrality by 2021. The government implemented policies to promote renewable energy, reforestation, and sustainable land use.[18] This political commitment was supported by robust leadership, which mobilised resources, engaged various stakeholders, and set clear targets. The success of this initiative demonstrates how strong political will and effective leadership can drive significant progress in sustainable development.

These elements are fundamental in ensuring that the chosen model is not only adopted but also actively supported and successfully implemented. Ultimately, the decision on which model to adopt often hinges on the political will of the country's leadership. Leaders who prioritise sustainability are more likely to champion comprehensive models, even if they require significant changes to existing practices. Hence, a country decides on the best model for sustainable development through a careful consideration of its unique circumstances,

18 Government of Costa Rica, *Costa Rica's Carbon Neutrality Goal* (San José: Government of Costa Rica, 2019), <https://www.cdn.go.cr/en/costa-ricas-carbon-neutrality-goal>, accessed 29 August 2024.

international obligations, stakeholder input, and an iterative process of testing and adapting different approaches. The choice is rarely about selecting a single model but rather about finding the right combination of principles that align with national goals and challenges.

Collaborative governance and its essential role in sustainable development governance

Collaborative governance is increasingly recognised as a crucial framework for achieving sustainable development. Sustainable development governance, which refers to the institutional processes and mechanisms used to integrate environmental, social, and economic objectives, requires the participation of diverse stakeholders. Collaborative governance, characterised by joint decision-making processes among governments, private sectors, civil society, and other stakeholders, is essential in ensuring that sustainable development goals (SDGs) are effectively implemented.

Integration of multiple stakeholders

One of the core reasons collaborative governance is essential for sustainable development is the necessity of integrating multiple stakeholders with varying interests and expertise. Sustainable development is inherently complex, involving a multitude of interrelated factors such as economic growth, environmental protection, and social equity. Collaborative governance allows for the inclusion of diverse perspectives and knowledge bases, which is vital in crafting policies that are comprehensive and balanced. By involving all relevant actors – governments, NGOs, businesses, and local communities – collaborative governance ensures that policies are more likely to be accepted and successfully implemented, as they reflect the collective input and needs of the society.

Enhanced accountability and transparency

Another significant benefit of collaborative governance in sustainable development is the enhancement of accountability and transparency. In traditional governance models, decisions are often made by a few central authorities, which can lead to a lack of transparency and accountability. Collaborative governance, in contrast, involves open dialogues and shared responsibilities among all stakeholders. This inclusiveness not only promotes transparency but also ensures that stakeholders hold each other accountable for the decisions made and actions taken. This level of accountability is crucial in the context of sustainable development, where long-term commitments and consistent policy implementation are required to achieve the desired outcomes.

Facilitating policy innovation and adaptive management

Collaborative governance also plays a pivotal role in fostering policy innovation and adaptive management, both of which are essential for sustainable development. The collaborative process encourages the sharing of ideas and experiences, which can lead to innovative solutions to complex sustainability challenges. Furthermore, collaborative governance allows for adaptive management, which is the ability to adjust policies and strategies in response to new information and changing circumstances. This flexibility is particularly important in sustainable development governance, where environmental, social and economic conditions are constantly evolving.

Building social capital and trust

Finally, collaborative governance contributes to building social capital and trust among stakeholders, which is essential for the long-term success of sustainable development initiatives. Trust is a fundamental element in any governance process, as it facilitates cooperation and reduces conflicts. Collaborative governance fosters a sense of ownership and commitment to the outcomes, thereby enhancing trust and cooperation by involving stakeholders in the decision-making process. This social capital is invaluable in achieving sustainable development goals, as it ensures that all parties are working towards a common vision and are willing to support each other in overcoming challenges. In conclusion, collaborative governance is indispensable for sustainable development governance. Its ability to integrate diverse stakeholders, enhance accountability and transparency, foster innovation and adaptive management and build social capital and trust makes it a vital approach in achieving sustainable development goals. As the world continues to face complex sustainability challenges, the importance of collaborative governance will only grow, necessitating its integration into all levels of sustainable development governance.

Conclusion

It is evident that translating theoretical models into practical solutions requires a multifaceted approach. The chapter has explored various models of sustainable development, highlighting their theoretical underpinnings and practical applications. The effectiveness of these models hinges on several key factors, including alignment with international commitments, stakeholder consultation, pilot projects, comparative analysis, policy coherence, flexibility, adaptability, monitoring, evaluation, and political will. Each model offers unique strengths and challenges, demonstrating that no single approach fits all contexts. Successful implementation often involves adapting models to specific national and local conditions, leveraging lessons learned from pilot projects and case studies, and ensuring robust monitoring and evaluation systems are in place. Political will and leadership play a crucial role in driving these models forward, providing the necessary support and resources. Bridging the gap between theory and practice requires a dynamic and inclusive approach, where theoretical models are continuously tested, refined, and adapted to real-world conditions. This iterative process ensures that sustainable development efforts are both effective and resilient, paving the way for meaningful progress towards global sustainability goals.

The journey from theory to practice in sustainable development models underscores the need for a nuanced understanding of local contexts and challenges. While theoretical models provide a framework for action, their real-world effectiveness is shaped by how well they are adapted and implemented. This chapter illustrates that practical success involves integrating diverse factors such as stakeholder input, flexible policy design, and ongoing evaluation. Moreover, it highlights the importance of political leadership in championing these models and ensuring they receive the necessary support. Embracing an adaptive, evidence-based approach allows for continuous improvement and responsiveness, ultimately driving more impactful and sustainable outcomes.

Role of International Agencies in Sustainable Development

Introduction

The first half of the twentieth century went into creating new permanent world organizations that would strive to maintain peace and security around the globe. Besides dealing with sovereignty issues, these organisations dealt with economic and social problems. A brief history of international organisations is vital to understanding how they developed in the second half of the twentieth century and how they play a significant role in implementing sustainable development goals worldwide. Following is the timeline to understand their emergence:

1814–1815 The Vienna Congress was convened to maintain peace and restore power balance.

1865 The International Telegraph Union was established.

1874 The International Railway Congresses Association was established.

1889 International NGOs (INGOs) multiplied in large numbers, such as the World Zionist Organization, International Statistical Institute, and International Actuarial Association.

1899-1907 Hague Conferences were held to keep the European family of states as an organised entity.

1919 After the First World War, the heads of state and government, foreign ministers and their advisers met at Versailles to sign the Peace Treaty to uphold sovereignty and peace with perfunctory reference to economic and social issues.

1920 The League of Nations was formed with headquarters in Geneva, Switzerland, to resolve international disputes. It provided support for socio-economic problems.

1945 The United Nations (UN) was established on June 26 1945 in San Francisco, California (USA) with 51 members. The League of Nations was dissolved in 1946.

1944 The World Bank Group was formed following the Bretton Woods agreements, with 38 members, increasing to 189. It works through five agencies.

1961 The OECD was established with 35 member countries, increasing to 38 to stimulate economic progress and world trade.

1961 Amnesty International was founded in London to research and act against severe human rights abuses, including violations of physical and mental integrity, freedom of conscience and expression and discrimination.

1966 The ADB was established as a regional development bank headquartered in Manila, Philippines, dedicated to reducing poverty in Asia and the Pacific through loans.

1967 The Association for Southeast Asian Nations (ASEAN) was established to foster political and economic cooperation and to ensure regional stability in Southeast Asia.

1971 United Nations Environment Programme (UNEP) was created to coordinate global environmental activities and assist countries in implementing environmentally sound policies.

1988 Intergovernmental Panel on Climate Change (IPCC) was established to assess the science related to climate change and its impacts.

1993 European Union (EU) was formed with the Maastricht Treaty, it represents European countries' political and economic union.

1995 World Trade Organization (WTO) – Established to regulate international trade and resolve trade disputes.

2000 Millennium Development Goals (MDGs) were launched by the United Nations to find solutions to global challenges such as poverty, hunger, disease, and education.

2015 Sustainable Development Goals (SDGs) were adopted by the United Nations to continue and expand upon the MDGs, aiming for a more comprehensive approach to sustainable development by 2030.

This timeline reflects the evolution and emergence of international organizations, highlighting their role in addressing global challenges and promoting cooperation among nations. There are around 77,761 operational and roughly 38,000 inactive international organizations representing 300 countries and territories, which include government international organizations and international non-governmental organizations (INGOs).[1] International organizations have emerged in response to the growing need to develop international relations to promote socioeconomic progress and maintain peace, harmony and sovereignty. Global trade grew after the Second World War, which required regulation via institutional means. Since the early times, world history has seen many organizations for trade amongst countries and faith organizations for countries dealing with social welfare issues. However, only after the 1950s did the world see a literal outburst in international organizations working to achieve socioeconomic goals.

This chapter will cover the major international organizations contributing to sustainable development.

United Nations

It is essential to understand the historical background and the organizational structure of the UN before its contributions to the developmental tasks are studied. The UN was established in June 1945 in San Francisco, USA. The Charter of the UN Conference on International Organization was signed and came into force on October 24 1945. The Statute of the International Court of Justice (ICJ) is integral to the Charter. The UN was formulated as a response to the League of Nations' failure to be effective. The number of its members increased from fifty one in 1945 to 193 at present. Its name, 'United Nations', was coined by the United States President Franklin D. Roosevelt and was first used in the Declaration by the UN representatives of twenty-six nations that pledged that their governments would fight together against the Axis Powers.[2]

Main bodies of the UN

The UN comprises six central bodies.

1 Union of International Associations, 'Yearbook of International Organizations', <https://uia.org/ybio/>, accessed 4 August 2024.

2 In 1939, under the 'Pact of Steel' with the Tripartite Pact of 1940, the Rome–Berlin Axis emerged, bringing together the military objectives of Germany, Italy, and Japan; the Axis Powers were defeated in 1945, and their alliance was dissolved; the Allied Powers or Entente Powers (members of the original Triple Entente of 1907: the French Republic, the British Empire, and the Russian Empire) were opposed to the Axis Powers; the Allied Powers were led by the US from 1941 to 1945.

1. The General Assembly: The General Assembly is one of the six organs of the UN, the only one in which all the member states have equal representation of one nation one vote. All the representatives of 193 member nations meet every September to discuss and work together on international issues covered by the UN Charter, such as development, peace, security and international law. The General Assembly elects its president for each session. Decisions on important issues such as peace, security, admission of new members and budgetary matters require a two-thirds majority. In contrast, decisions on other questions are taken by a simple majority.

Functions and powers of the General Assembly

The General Assembly can recommend international issues to states within its competence. It has carried out initiatives that affect the political, economic, humanitarian, social and legal aspects of the lives of millions around the world. It adopted the Millennium Declaration in 2000 and the 2005 Summit Outcome, which focused on development, poverty eradication, safeguarding human rights, promoting the rule of law, and environmental protection. In September 2015, the General Assembly agreed on seventeen SDGs as a post-2015 development agenda to be achieved by 2030.

Functions of the General Assembly

The following are the primary functions of the General Assembly:

- Elect non-permanent members
- Approve the UN budget
- Elect the members of other United Nations councils and organs, and on the recommendation of the Security Council, appoint the Secretary-General. Make recommendations on general principles of cooperation for maintaining international peace and security, including disarmament
- Settle international disputes except when the Security Council is discussing it

General Assembly and Sustainable Development

The United Nations General Assembly plays a crucial role in sustainable development by setting global agendas, adopting frameworks and policies, and mobilizing resources to address critical issues. Here are some key aspects of how the General Assembly contributes to sustainable development:

- Adopting global agendas: The General Assembly adopts major international frameworks that guide sustainable development efforts. The most notable example is adopting the 2030 Agenda for Sustainable Development, which includes the 17 Sustainable Development Goals (SDGs). This agenda provides a comprehensive plan for achieving sustainable economic, social, and environmental development.[3]

- Facilitating international cooperation: The General Assembly is a platform for member states to discuss and collaborate on sustainable development issues. It encourages international cooperation and partnerships among countries, international organizations, the private sector and civil society through resolutions and declarations.[4]
- Monitoring and reviewing progress: The General Assembly monitors and reviews the implementation of sustainable development goals. It holds High-Level Political Forums (HLPF) on sustainable development, which provide a space for member states to present their progress, share best practices, and address challenges.[5]
- Mobilizing resources: The General Assembly mobilises financial and technical resources necessary for sustainable development. It advocates for increased funding, technology transfer and capacity-building support to help countries implement the SDGs.[6]
- Raising awareness and advocacy: The General Assembly raises awareness about sustainable development issues and advocates for global action. It organises events, thematic debates and campaigns to highlight the importance of sustainable development and the need for urgent action.[7]

2. The Security Council: The Security Council is responsible for maintaining international peace and security. It comprises fifteen members. Wach has one vote, and all member states must comply with its decisions. The Security Council works to determine the existence of a threat to peace or an act of aggression. It calls on the parties for dispute resolution and can resort to sanctions or authoritarian means, such as using force to restore international peace. The Security Council meets whenever peace is

3 United Nations, 'Transforming our world: The 2030 agenda for sustainable development', (New York: United Nations, 2015), <https://sustainabledevelopment.un.org/post2015/transformingourworld>, accessed 3 August 2024.

4 United Nations General Assembly, *Resolutions and Decisions Adopted by the General Assembly during its 71st Session* (New York: United Nations, 2017).

5 United Nations General Assembly, *High-Level Political Forum on Sustainable Development* (New York: United Nations, 2023).

6 United Nations General Assembly, *Financing for Development: Progress and Prospects* (New York: United Nations, 2023).

7 United Nations General Assembly, *Sustainable Development Goals* (New York: United Nations, 2023).

threatened. The decisions of the Security Council, which has the authority to make enforceable decisions, must be accepted and implemented by all UN members according to the Charter.

Organization

The Security Council first met in 1946 at the Church House, Westminster, London. After that, it had its headquarters in New York City. It has also held sessions in Addis Ababa, Ethiopia (1972), Panama City, Panama (1973) and in Geneva, Switzerland (1990). A representative of each of its members must be present at all times at the UN Headquarters so that the Security Council can meet at any time the need arises.

UNSC and sustainable development

The United Nations Security Council (UNSC) plays a significant role in sustainable development through its mandate to maintain international peace and security. The linkage between peace, security and development is widely recognised, as sustainable development cannot be achieved without a stable and secure environment. The UNSC creates conditions conducive to sustainable development by addressing and resolving conflicts. Armed conflicts can devastate economies, displace populations, and destroy infrastructure, hindering development efforts. UNSC-authorised peacekeeping missions help stabilise regions, allowing for the rebuilding of communities and economies. These missions often include components that support establishing governance structures, the rule of law and protecting human rights, which are essential for sustainable development (United Nations Security Council, 'The Role of the Security Council in Promoting Sustainable Development'.

The UNSC addresses issues related to the exploitation of natural resources, which can fuel conflicts by finding ways to manage natural resources responsibly. The UNSC helps ensure these resources contribute to sustainable development rather than conflict. The UNSC is critical in post-conflict reconstruction efforts, which are vital for sustainable development. This includes supporting the rebuilding of infrastructure, the return of displaced persons and the re-establishment of governance and economic systems. The UNSC also deals with non-traditional security threats that impact sustainable development, such as climate change, terrorism, and pandemics. The UNSC helps create a secure environment necessary for sustainable development by addressing these threats.[8]

---

8 Nalin Desai, 'Peacekeeping and sustainable development: How UN missions contribute to development goals', *Journal of International Peace Operations* 12.3 (2021): 15–30.

3 UN ECOSOC

The leading role of ECOSOC is to advance the three dimensions of sustainable development: economic, social and environmental. It promotes innovative thinking, debates, finding consensus on ways forward and coordinating the efforts to achieve internationally agreed goals.

Composition

The ECOSOC comprises fifty-four members of the UN who are elected by the General Assembly, keeping in mind the geographical representation, for an overlapping three-year term.

Functions

The ECOSOC operates at the centre of the UN system's work on sustainable development, covering the economic, social and environmental aspects. Its basic functions are the following:

- Promote sustainable development
- Manage the transitions from MDGs to SDGs
- Advance policy integration
- Finance for sustainable development
- Coordinate humanitarian action
- Guide operational activities for development
- Provide coordination and oversight
- Build partnerships
- Engage youth
- Raise awareness of emerging issues
- Advise on Haiti's long-term development
- Work with CS

Sustainable Development

ECOSOC undertakes varied initiatives to promote SDGs. Some of them are:

- High-level political forum on sustainable development
- UN Conferences and high-level events on sustainable development
- Multi-stakeholder Forum on Science, Technology and Innovation on SDGs

The Sustainable Development Goals Report 2024 outlines the world's major obstacles in making significant progress toward achieving the SDGs based on the latest data and projections. It identifies areas with setbacks but highlights regions where notable progress has been made, such as reducing global child mortality, preventing HIV infection and improving access to energy and mobile broadband. The report underscores the need for accelerated action, particularly in critical areas hindering SDG progress, including climate change, peace and security, and inequalities within and between countries. With only six years remaining, the report indicates that current progress is insufficient to meet the SDGs. Without substantial investment and increased efforts, achieving the SDGs, which is essential for a more resilient and prosperous world and as a solution to current global crises, will remain out of reach. The lingering effects of the COVID-19 pandemic, rising conflicts, geopolitical tensions and increasing climate disruptions have significantly impeded progress. The report emphasises urgent priorities and areas where more robust and effective actions are needed to fulfil the 2030 promise to eradicate poverty, protect the planet, and ensure no one is left behind.[9]

4 Trusteeship Council

The Trusteeship Council was created as a vital part of the UN in 1945 as part of the International Trusteeship System to supervise freeing dependent territories from colonial powers. This was established during the Chapter 12 discussions at the San Francisco Conference under the Charter of the UN.

Composition

The Trusteeship Council comprises five permanent members of the Security Council: China, France, the Russian Federation, the United Kingdom and the United States of America.

Functions and powers

The main goal of the Trusteeship Council was to promote the advancement of the inhabitants of the trust territories and their progressive development towards self-government or independence. It comprises an elected president and vice president for five years.

Status

The Trusteeship Council suspended its operations on November 1 1994, after it had gained independence from Palau, the last remaining UN Trust Territory. By a resolution adopted on May 25 1994, the Trusteeship Council amended its rules of procedure to drop the obligation to meet annually and agreed to meet as the occasion required, by the decision or at the request of a majority of its

9 United Nations, 'The sustainable development goals report 2024', <https://unstats.un.org/sdgs/report/2024>, accessed 6 August 2024.

members or the General Assembly or the Security Council. Although its direct role in sustainable development has been minimal, the Trusteeship Council's legacy and principles, like self-determination, political independence and social progress, align with sustainable development goals.

5 International Court of Justice

The ICJ is the principal judicial organization of the UN, established in 1945 by the UN Charter under Article 33 and began its work in April 1946. Its headquarters is at the Peace Palace in The Hague (Netherlands). Its role is to settle, per international law, legal disputes and give advisory opinions on legal questions referred to by authorised UN organizations and specialised agencies.

Composition

The ICJ is comprised of fifteen judges who are elected for a term of nine years by the UN General Assembly and the Security Council. Both organizations vote simultaneously but separately in New York. In order to be elected, a candidate must receive an absolute majority of the votes in both bodies. One-third of the court members are elected every three years to ensure continuity. It is assisted by a Registry to handle its administrative affairs. Its official languages are English and French.

Contentious cases

The court entertains two types of cases: contentious cases and advisory proceedings. The legal disputes between the states that are submitted to the court are called contentious cases on the condition that the states must be parties to the Statute of the court or must have accepted its jurisdiction. The court discharges its duties according to the procedure laid down as a full court, but at the parties' request, it may also establish ad hoc chambers to examine specific cases.

Sources of law

The sources of law that the ICJ must draw from to decide the cases are as follows:

- International treaties and conventions in force
- International custom
- The general principles of law
- Judicial decisions
- Teachings of the most high-quality publicists

If the parties agree, the court can decide a case, *ex aequo et bono,* without limiting itself to the existing rules of international law.

Advisory proceedings

The requests for advisory opinions on legal questions referred to by the UN organs and specialised agencies are termed advisory proceedings. They are open solely to the five organs of the UN and to the sixteen specialised agencies of the UN family or affiliated organisations. The court is empowered to hold written and oral proceedings, which are certain aspects that recall the proceedings in contentious cases. Its advice is held as if it were sanctioned by international law.

Financial assistance to parties

The Secretary General's Trust Fund assists the states in settling disputes through the ICJ. In 1989, the UN Secretary-General set up a Trust Fund to provide the states with financial assistance in certain circumstances. At present, the Fund is open to all states where the jurisdiction of the court (or admissibility of the application) is not or is no longer the subject of dispute on its part. The Fund caters to helping the state parties to a dispute to comply with the judgement of the Court.

6 ICJ and Sustainable Development

The International Court of Justice (ICJ), as the principal judicial organ of the United Nations, contributes to sustainable development in several ways:

- Adjudication of disputes: The ICJ helps resolve disputes related to natural resources, environmental protection, and transboundary harm between states. The ICJ promotes sustainable management and use of resources by providing a legal framework for addressing these issues.
- Development of international environmental law: The ICJ contributes to developing and clarifying international environmental law through its rulings and advisory opinions. These decisions often set essential precedents for sustainable development, helping to establish norms and standards that guide state behaviour.
- Peace and security: Sustainable development is closely linked to peace and security. The ICJ contributes to a stable and secure international environment, essential for sustainable development by adjudicating disputes peacefully and promoting international law.
- Advisory opinions: The ICJ provides advisory opinions on legal questions referred to it by UN organs and specialised agencies. These opinions can influence international policies and practices related to sustainable development, including environmental protection, human rights and economic growth.
- Enforcement of international agreements: The ICJ ensures that states adhere to their international obligations, including environmental treaties

and sustainable development agreements. This helps maintain the integrity of international law and encourages states to commit to sustainable practices.

- Capacity building and legal assistance: The ICJ helps states develop their legal systems and institutions by providing legal assistance and capacity-building support. This enhances their ability to implement sustainable development policies and comply with international legal standards.

Secretariat

The Secretariat is one of the six principal organs of the UN, with each department or office having a distinct area of action and responsibility. The departments coordinate to ensure cohesion in the organisation's day-to-day work in offices and duty stations worldwide. The Secretary-General is the head of the UN Secretariat. Its headquarters is in New York.

Various offices and departments of the Secretariat are as follows:

- Executive Office of the Secretary General
- Office of Internal Oversight Services
- Office of the Legal Affairs
- Department of Political Affairs
- Office for Disarmament Affairs
- Department of Peacekeeping Operations
- Department of Field Support

The World Bank (WB)

The WB was conceived in 1944 at the Bretton Woods Monetary Conference in Bretton Woods, New Hampshire, initially aiming to help rebuild the European countries devastated by the Second World War. It later spread its ambit to the rest of the world to fund large infrastructure projects such as dams, electrical grids, irrigation systems and roads. The lending was not only of a financial nature but also of technical assistance and training to use the loans effectively.

Organisation

The WB is like a cooperative of 189 member countries and is represented by a board of governors, the ultimate policymakers at the WB. Generally, the governors are finance ministers or development ministers of member countries. They meet once a year at the annual meetings. There are 25 executive directors who work on-site at the Bank. The five largest shareholders appoint an executive director, while executive directors represent other member countries. Its headquarters are in Washington, DC, USA.

The World Bank Group consists of the following five organs:

- The IBRD: Lends to governments of middle-income and creditworthy low-income countries.
- The International Development Association: Provides internal free loans, called credits, and grants to the governments of the poorest countries.
- The International Financial Corporation: Helps developing countries achieve sustainable growth by financing investment, mobilizing capital in international financial markets and providing advisory services to private sector businesses and governments.
- The Multilateral Investment Guarantee Agency: Created in 1988 to promote FDI in developing countries to support economic growth, reduce poverty and improve people's lives. It also offers political risk insurance to investors and lenders.
- The International Centre for Settlement of Investment Disputes: Provides international facilities for conciliation and arbitration of investment disputes.

World Bank and sustainable development

The World Bank plays a significant role in promoting sustainable development worldwide. Here are several vital ways it contributes in brief:

Financial assistance

- Loans and grants: The World Bank provides financial resources to developing countries through loans and grants. These funds support projects to reduce poverty, improve infrastructure and foster sustainable economic growth.
- Concessional financing: The International Development Association (IDA), part of the World Bank Group, offers concessional financing to the world's poorest countries, focusing on projects that promote sustainable development.

Project implementation and support

- Infrastructure development: The World Bank funds projects in transportation, energy, water supply, and sanitation, which are critical for sustainable development. These projects aim to create a resilient infrastructure that can withstand environmental and economic challenges.
- Agriculture and rural development: The World Bank helps improve food security, increase productivity, and promote sustainable farming practices by supporting agricultural projects.

Environmental sustainability

- Climate change mitigation and adaptation: The World Bank finances projects that address climate change by reducing greenhouse gas emissions and helping countries adapt to the impacts of climate change. This includes investments in renewable energy, energy efficiency, and climate-resilient infrastructure.

- Natural resource management: The Bank supports initiatives to manage natural resources sustainably, including forests, water, and fisheries. These projects aim to balance economic development with environmental conservation.

Social development

- Education and health: The World Bank invests in education and healthcare to improve human capital, essential for sustainable development. These projects focus on increasing access to quality education and healthcare services, particularly for vulnerable populations.
- Social protection: The World Bank helps countries build systems to protect their most vulnerable citizens, ensuring that economic growth is inclusive and benefits all segments of society by funding social protection programs.

Capacity building and knowledge sharing

- Technical assistance and capacity building: The World Bank provides technical and capacity-building support to help countries strengthen their institutions and governance systems. This support is crucial for implementing and sustaining development projects.
- Research and data: The World Bank researches and publishes data on various aspects of sustainable development. Its reports and publications provide valuable insights and best practices that countries can use to inform their development strategies.

Promoting private sector development

- Private sector engagement: The International Finance Corporation (IFC), a member of the World Bank Group, promotes private sector development by investing in businesses and providing advisory services. The IFC supports projects that create jobs, enhance economic growth, and promote sustainable practices in the private sector.

Partnerships and collaboration

- Global partnerships: The World Bank collaborates with governments, international organizations, civil society, and the private sector to address global challenges and promote sustainable development. These partnerships leverage additional resources and expertise to achieve development goals.
- Multilateral initiatives: The World Bank participates in multilateral initiatives and frameworks, such as the Sustainable Development Goals (SDGs), to coordinate efforts and maximise the impact of development activities.

Thus, the World Bank supports sustainable development through financial assistance, project implementation, environmental sustainability initiatives, social development programs, capacity building, private sector engagement, and global partnerships. Its comprehensive approach addresses sustainable development's economic, social and environmental dimensions.

Food and Agriculture Organization (FAO)

The United Nations Food and Agriculture Organization (FAO) is a specialised agency founded in 1945 to defeat hunger and improve nutrition and food security worldwide. The FAO's headquarters is in Rome, Italy, and it operates globally in over 130 countries.

Functions

1. Information and data collection:
   - Global Statistics: The FAO collects, analyses and disseminates data on agriculture, forestry, fisheries, and food security. This information helps countries make informed decisions about their food and agricultural policies.
   - Publications: The FAO produces numerous reports, such as the annual State of Food Security and Nutrition in the World (SOFI), providing critical insights into global food security trends.
2. Policy and technical assistance:
   - Policy support: The FAO advises governments on agricultural policies, helping them develop strategies to improve food security, nutrition, and sustainable farming practices.
   - Capacity building: The FAO provides technical assistance and capacity-building programs to strengthen national and regional institutions involved in agriculture and food security.
3. Sustainable agricultural development:
   - Innovation and Research: The FAO promotes agricultural research and the adoption of new technologies to increase productivity while ensuring environmental sustainability.
   - Climate-Smart Agriculture: The FAO advocates for climate-smart agricultural practices that enhance resilience to climate change and reduce greenhouse gas emissions.
4. Emergency response and rehabilitation:
   - Disaster response: The FAO provides emergency assistance to countries affected by natural disasters, conflicts, and food crises, helping them restore agricultural production and rebuild livelihoods.

   - Resilience building: The FAO works on long-term recovery and resilience-building projects to help communities withstand future shocks.
5. Standard setting and regulation:
   - Codex Alimentarius: The FAO, in collaboration with the World Health Organization (WHO), establishes international food standards, guidelines and codes of practice through the Codex Alimentarius Commission to ensure food safety and quality.
   - Phytosanitary Standards: The FAO develops international standards for plant health through the International Plant Protection Convention (IPPC) to protect plants from pests and diseases.

Role in sustainable development

The FAO plays a crucial role in promoting sustainable development, mainly through the following areas:

1. Sustainable agriculture and food systems:
   - Agroecology: The FAO promotes agroecological approaches that enhance biodiversity, improve soil health, and reduce dependency on chemical inputs, contributing to sustainable agricultural systems.
   - Integrated Pest Management (IPM): The FAO supports IPM practices that minimise harmful pesticides and promote natural pest control methods.
2. Food security and nutrition:
   - Zero hunger goal: The FAO is a crucial player in the global effort to achieve Sustainable Development Goal 2 (SDG 2), which aims to end hunger, achieve food security, improve nutrition and promote sustainable agriculture by 2030.
   - Nutrition improvement: The FAO improves diets and nutrition through programs promoting diversified and healthy diets, particularly in vulnerable populations.
3. Climate change mitigation and adaptation:
   - Sustainable land and water management: The FAO implements projects that promote sustainable land use and water management practices to combat desertification, soil degradation and water scarcity.
   - Climate resilience: The FAO supports countries in developing climate-resilient agricultural practices and policies, helping farmers adapt to changing climate conditions.
4. Biodiversity conservation:
   - Genetic resources: The FAO works to conserve and sustainably use plant and animal genetic resources, ensuring the preservation of biodiversity crucial for food security and ecosystem health.

  - Forestry and fisheries: The FAO promotes sustainable management of forests and fisheries, protecting these vital resources from overexploitation and degradation.
5. Economic and social development:
  - Rural development: The FAO focuses on improving the livelihoods of rural communities, reducing poverty, and enhancing economic opportunities through sustainable agricultural development.
  - Women's empowerment: The FAO advocates for gender equality and women's empowerment in agriculture, recognizing their critical role in food production and rural economies.

Hence, the FAO is pivotal in advancing sustainable development by promoting sustainable agriculture, ensuring food security and nutrition, enhancing climate resilience, conserving biodiversity, and supporting economic and social development. Its comprehensive approach and global reach make it a key player in addressing the complex challenges of sustainable development.[10]

International Monetary Fund (IFC)

The IMF is an organization of 189 countries working to foster global monetary cooperation, secure financial stability, facilitate international trade, promote high employment and sustainable economic growth, and reduce poverty worldwide. It was created in 1945 and is governed by and accountable to 189 countries that comprise its near-global membership. Its headquarters are in Washington, DC.

India joined the IMF on December 27, 1945, as one of the original members of the IMF. India accepted the obligations of Article VIII of the IMF Articles of Agreement on current account convertibility on August 20 1994. India subscribed to the IMF's Special Data Dissemination Standard. Countries in this group commit to observing the standard and providing information about their data and data dissemination practices.

Composition

The Board of Governors, the highest decision-making body of the IMF, consists of one governor and one alternate governor for each member country. The member country appoints the governor, usually the minister of finance or the governor of

10 Food and Agriculture Organization of the United Nations, 'Sustainable development goals helpdesk', <https://www.fao.org/sustainable-development-goals-helpdesk/en>, accessed 4 August 2024.

the Central Bank. All powers of the IMF are vested in the Board of Governors. They may delegate to the Executive Board, all except certain powers. The Board typically meets once a year. India has a quota of Special Drawing Right (SDR) of 415,820 million and has no outstanding loans. The last consultation of the Board was on January 25 2017.

IMF's role in sustainable development

While the IMF's primary mandate focuses on economic stability and growth, its activities significantly contribute to sustainable development through various channels:

1. Promoting macroeconomic stability
   - Economic stability as a foundation for sustainable development: The IMF helps countries achieve macroeconomic stability, which is essential for sustainable development. Stable economies are better positioned to invest in social and environmental programs that support sustainable development goals.
   - Crisis prevention and resolution: By providing policy advice, financial support and technical assistance, the IMF helps countries prevent and resolve economic crises, which can devastate poverty and development.
2. Supporting inclusive growth
   - Poverty reduction and inclusive growth: The IMF's programs often include measures to protect vulnerable populations and promote inclusive growth. The IMF contributes to reducing poverty and inequality by advising on policies that foster job creation, equitable income distribution, and social safety nets.
   - Social spending and safety nets: The IMF encourages member countries to prioritise social spending, including healthcare, education and social protection programs, which are vital for sustainable development.
3. Addressing climate change
   - Climate-related financial policies: The IMF recognises the economic implications of climate change and provides policy advice on climate-related financial risks. This includes promoting carbon pricing, phasing out fossil fuel subsidies, and encouraging green investments.
   - Climate resilience and adaptation: The IMF supports countries in building resilience to climate change by integrating climate risks into economic planning and promoting adaptation strategies.
4. Enhancing governance and institutions

- Good governance: The IMF promotes good governance and strong institutions, which are critical for sustainable development. This includes combating corruption, improving transparency and strengthening public financial management.
- Institutional capacity building: Through technical assistance and training, the IMF helps countries build the capacity of their institutions to design and implement effective policies for sustainable development.

5. Supporting sustainable debt management
   - Debt sustainability: The IMF assists countries in managing their debt sustainably, ensuring that they do not accumulate unsustainable levels of debt that could jeopardise their development prospects. This includes debt relief initiatives for the heavily indebted economically disadvantaged.
   - Debt transparency and accountability: The IMF advocates for debt transparency and accountability, which helps countries maintain sustainable debt levels and fosters trust with international creditors.
6. Promoting global cooperation
   - International cooperation and policy coordination: The IMF facilitates international cooperation on economic and financial issues, promoting policy coordination among its member countries. This cooperation is essential for addressing global challenges such as climate change, financial instability and economic inequality.
   - Support for multilateral initiatives: The IMF collaborates with other international organizations and supports multilateral initiatives that align with sustainable development goals, such as the United Nations' 2030 Agenda for Sustainable Development.

Hence, the IMF contributes to sustainable development by promoting macroeconomic stability, supporting inclusive growth, addressing climate change, enhancing governance and institutions, supporting sustainable debt management, and promoting global cooperation. Through these efforts, the IMF helps create the economic conditions necessary for achieving the Sustainable Development Goals (SDGs).[11]

World Health Organization

WHO came into existence when its constitution was enacted on April 7 1948. At present, it has 7,000 employees working in 150 country offices. Its headquarters

11 International Monetary Fund, 'IMF and the Sustainable Development Goals', <https://www.imf.org/en/About/Factsheets/Sheets/2023/IMF-Sustainable-development-goals-SDGs>, accessed 4 August 2024.

is in Geneva. All member states' delegates attend the World Health Assembly and determine the organisation's policies. The Executive Board is composed of members technically qualified in health and affects the decisions and policies of the Health Assembly. WHO's core function is to direct and coordinate international health work through collaboration. It partners with countries, the United Nations system, international organizations, civil society, foundations, academia and research institutions.

Core Functions of WHO

The WHO's Secretariat focuses its work on the following six core functions in India:

- Providing leadership on matters critical to health and engaging in partnerships where joint action is needed.
- Shaping the research agenda and stimulating the generation, translation and dissemination of valuable knowledge.
- Setting up norms and standards and promoting and monitoring their implementation.
- Articulating ethical and evidence-based policy options.
- Providing technical support, catalysing change and building sustainable institutional capacity.
- Monitoring the health situation and assessing health trends.[12]

WHO and sustainable development

The WHO is crucial in advancing sustainable development, significantly through its contributions to the health-related Sustainable Development Goals (SDGs). Here's how the WHO supports sustainable development:

1. Promoting Universal Health Coverage (UHC)
   - Health for all: The WHO advocates for universal health coverage, ensuring everyone can access the health services they need without suffering financial hardship. UHC is a key target under SDG 3 (good health and well-being).
   - Strengthening health systems: The WHO supports countries in building resilient and efficient health systems that can deliver quality health services equitably.
2. Addressing health determinants

12 World Health Organization, 'About WHO', <https://www.who.int/about>, accessed 4 August 2024.

- Social determinants of health: The WHO addresses the social determinants of health, such as education, employment, housing and nutrition, which are crucial for achieving broader sustainable development.
- Environmental health: The WHO works on environmental health issues, including air and water quality, sanitation and chemical safety, directly impacting public health and sustainable development.

3. Tackling communicable and non-communicable diseases
   - Disease prevention and control: The WHO supports global efforts to prevent and control communicable diseases like HIV/AIDS, tuberculosis, malaria and, more recently, COVID-19. Controlling these diseases is essential for sustainable development as they significantly impact economic productivity and social stability.
   - Non-Communicable Diseases (NCDs): The WHO addresses the growing burden of NCDs, such as heart disease, stroke, cancer, diabetes, and chronic respiratory diseases. The WHO contributes to populations' long-term health and productivity by promoting healthy lifestyles and reducing risk factors.
4. Maternal, child and adolescent health
   - Improving health outcomes: The WHO works to improve health outcomes for mothers, children, and adolescents through initiatives that enhance access to quality healthcare, nutrition and education.
   - Reducing mortality: The WHO supports efforts to reduce maternal and child mortality rates, aligning with SDG targets to ensure healthy lives for all.
5. Emergency preparedness and response
   - Health emergencies: The WHO coordinates international responses to health emergencies, including disease outbreaks, natural disasters and conflicts. Effective emergency response is critical for sustainable development as it mitigates the health impacts of crises and supports recovery efforts.
   - Building resilience: The WHO helps countries build resilience to health emergencies by strengthening health systems and promoting preparedness and response capacities.
6. Global health governance and advocacy
   - Health policies and standards: The WHO sets international health policies and standards, ensuring worldwide consistency and quality in health practices.
   - Advocacy and partnerships: The WHO advocates for global health priorities and collaborates with governments, international organizations, civil society and the private sector to achieve health and development goals.

7. Specific initiatives and programs
   - Global Action Plan for Healthy Lives and Well-being for All: The WHO leads this initiative, which brings together twelve global health organizations to accelerate progress towards the health-related SDGs.
   - Framework Convention on Tobacco Control (FCTC): The WHO's FCTC is a global public health treaty aimed at reducing the harmful effects of tobacco consumption, a significant risk factor for NCDs.
   - WHO Health Emergencies Programme: This program enhances the capacity of countries and communities to manage health emergencies and provides rapid response to crises.

Thus, the WHO plays a pivotal role in sustainable development by promoting universal health coverage, addressing social and environmental determinants of health, tackling communicable and non-communicable diseases, improving maternal and child health, responding to health emergencies and advocating for global health governance. Through these efforts, the WHO supports the achievement of the Sustainable Development Goals and contributes to healthier, more resilient and equitable societies.[13]

United Nations International Children's Emergency Fund

UNICEF advocates for protecting children's rights to help meet their basic needs and expand their opportunities to reach their full potential. It is committed to ensuring special protection for the most disadvantaged children, whether victims of war, disasters, extreme poverty, violence and exploitation and those with disabilities.

UNICEF was founded on December 11, 1946, in New York City, New York, USA. Its headquarters are in New York, under its parent organization, UNGA. Initially, it was formed to provide emergency food and health care to children in countries devastated by the Second World War. It is funded by voluntary contributions from governments, NGOs, foundations, corporations and private individuals, and it receives no funding from the assessed dues of the UN. It has a presence in 190 countries.

13 World Health Organization Europe, 'Sustainable Development Goals', <https://www.who.int/europe/about-us/our-work/sustainable-development-goals>, accessed 4 August 2024.

Executive board

Its executive board is the governing body of UNICEF, providing intergovernmental support and oversight to the organization. It comprises thirty-six members representing the UN's five regional groups of member states. The Bureau coordinates its work, and the Executive Board must meet thrice each calendar year. Each country office carries out UNICEF's mission through a unique programme of cooperation developed with the host government for a five-year programme as a part of other UN activities in a country.

UNICEF and sustainable development

UNICEF plays a vital role in sustainable development by focusing on children's well-being, which is central to the Sustainable Development Goals (SDGs). Here's how UNICEF contributes to sustainable development:

1. Health and well-being (SDG 3)
    - Immunisation: UNICEF is the world's largest provider of vaccines to developing countries, helping to immunise millions of children each year against preventable diseases.
    - Nutrition: UNICEF implements programs to address malnutrition, promote breastfeeding, provide micronutrient supplements, and treat severe acute malnutrition.
    - Maternal and child health: UNICEF supports maternal health programs to reduce maternal mortality and improve neonatal health.
2. Quality education (SDG 4)
    - Access to education: UNICEF works to ensure that all children, particularly girls and marginalised children, have access to quality education. This includes building schools, training teachers and providing learning materials.
    - Early childhood development: UNICEF supports early childhood education programs that provide a strong foundation for lifelong learning and development.
    - Education in emergencies: UNICEF provides educational opportunities for children affected by emergencies, ensuring continuity of learning during crises.
3. Gender equality (SDG 5)
    - Empowering girls: UNICEF promotes gender equality by advocating for girls' education, preventing child marriage, and supporting programs that empower adolescent girls.

- Ending violence against women and girls: UNICEF works to eliminate violence against women and girls, including harmful practices like female genital mutilation (FGM) and child marriage.

4. Clean water and sanitation (SDG 6)
   - Water, sanitation, and hygiene (WASH): UNICEF provides clean water, sanitation facilities, and hygiene education to communities, reducing waterborne diseases and improving health outcomes.
5. Reducing inequality (SDG 10)
   - Inclusive policies: UNICEF advocates for inclusive policies that ensure all children, including those from marginalised and vulnerable groups, have equal opportunities to thrive.
   - Social protection: UNICEF supports social protection programs that provide financial assistance and services to disadvantaged families.
6. Climate action (SDG 13)
   - Climate resilience: UNICEF helps communities build resilience to climate change by supporting climate-smart agriculture, promoting clean energy solutions and raising awareness about climate impacts on children.
   - Disaster risk reduction: UNICEF works to reduce the risks of disasters by supporting preparedness and response efforts that prioritise children's needs.
7. Partnerships for the goals (SDG 17)
   - Global partnerships: UNICEF collaborates with governments, international organizations, civil society and the private sector to advance the SDGs and improve the lives of children globally.
   - Advocacy and Mobilization: UNICEF mobilises resources and advocates for increased investment in children's health, education and protection.

Specific Initiatives and Programs

- Global Polio Eradication Initiative: UNICEF is a key partner in this initiative, working to eradicate polio through vaccination campaigns and public health education.
- No Lost Generation Initiative: This program provides education, protection, and opportunities for children and youth affected by the Syrian crisis.
- Every Child Alive Campaign: UNICEF's campaign to end preventable newborn deaths by ensuring that every child has access to affordable, quality healthcare.

UNICEF's efforts in health, education, gender equality, water and sanitation, reducing inequality, climate action, and global partnerships are integral to sustainable development. By focusing on the rights and needs of children, UNICEF ensures that the next generation can contribute to a more equitable, sustainable, and prosperous world.[14]

UNESCO

UNESCO supports countries in developing and expanding educational activities focusing on sustainability issues such as climate change, biodiversity, disaster risk reduction, water, the oceans, sustainable urbanization and sustainable lifestyles through ESD. UNESCO leads and advocates globally on ESD and provides guidance and standards. It also provides data on the status of ESD and monitors progress on SDG Indicator 4.7.1 on the extent to which global citizenship education and ESD are mainstreamed in national education policies, curricula, teacher education and student assessment.[15]

UNESCO aids countries in developing and enhancing educational initiatives that address sustainability issues like climate change, biodiversity, disaster risk reduction, water, oceans, sustainable urbanization and sustainable lifestyles through Education for Sustainable Development (ESD). As a global leader and advocate for ESD, UNESCO offers guidance and standards, collects data on ESD status, and monitors progress on SDG Indicator 4.7.1. This indicator measures global citizenship education and ESD integration into national education policies, curricula, teacher education, and student assessment.

UNESCO and sustainable development

- Education for Sustainable Development (ESD): UNESCO provides educational programs and resources to help individuals understand and tackle sustainable development's environmental, social and economic challenges.
- Heritage preservation: UNESCO is dedicated to preserving cultural and natural heritage sites, including World Heritage sites, to promote sustainable tourism and cultural exchange.
- Scientific cooperation: UNESCO fosters scientific research and collaboration to support sustainable development and address global environmental issues like climate change and biodiversity loss.
- Media and information literacy: UNESCO aims to improve media and information literacy, empowering people to engage in sustainable development discussions and make informed decisions.

---

14 Adapted from: United Nations Children's Fund, 'About UNICEF', <https://www.unicef.org>, accessed 4 August 2024.

15 Adapted from: United Nations Educational, Scientific and Cultural Organization, 'Global Citizenship and Peace Education', <https://www.unesco.org/en/global-citizenship-peace-education>, accessed 4 August 2024.

- Cultural diversity: UNESCO advocates for cultural diversity as a core element of sustainable development, recognizing the importance of cultural expression and heritage in fostering social cohesion and a sense of identity.
- Gender equality: UNESCO acknowledges gender equality as a crucial component of sustainable development and works to eradicate gender-based discrimination and empower women and girls.[16]

The Organization for Economic Cooperation and Development (OECD)

OECD was established on September 30 1961. Its history began in 1948 when the Organization for European Economic Cooperation (OEEC) was established in 1948 to run the US-financed Marshall Plan to reconstruct a war-ravaged continent. Canada and the United States of America joined the OEEC after its success was established and the new OECD Convention was signed on December 14 1960, which came into force in 1961. Currently, it has thirty-five member countries that work together to identify, discuss and analyse each other's problems. It has four non-members, which makes the total number of members thirty-nine, accounting for 80 per cent of world trade and investment.

Composition

The OECD is comprised of a council at the top, with various committees to carry out its activities and work with the Secretariat, as shown in Figure 22.

- Council: The decision-making power is vested in the OECD Council. It is made up of one representative per member country plus a representative of the European Commission. The Council meets regularly, and the OECD

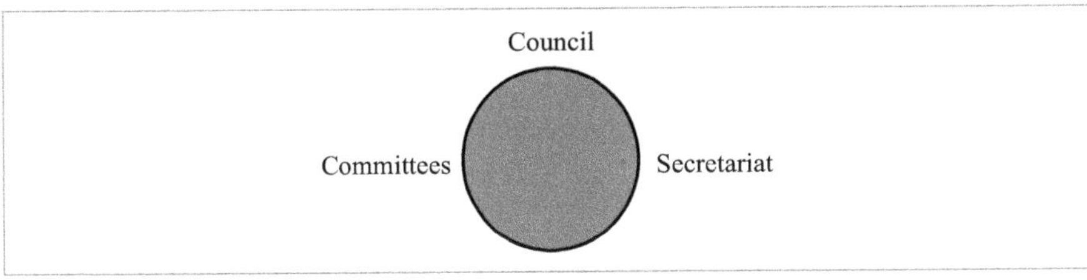

Figure 22 Composition of OECD

16 Adapted from: United Nations Educational, Scientific and Cultural Organization, 'Sustainable development', <https://www.unesco.org/en/sustainable-development>, accessed 4 August 2024.

Secretary-General chairs these meetings. The work mandated by the Council is carried out by OECD Secretariat.[15]

- Committees: Representatives of the thirty-five OECD member countries meet in the specialised committees to advance ideas and review progress in specific policy areas such as economics, trade, science, employment, education or financial markets. There are 250 Committees, working groups and expert groups.
- Secretariat: Angel Gurría heads the OECD Secretariat and is assisted by one or more deputy Secretaries providing the link between national delegations and the Secretariat.

Functions

OECD focuses on helping governments around the world. Its main functions are to do the following:

- Restore confidence in markets and the institutions that make them function.
- Re-establish healthy public finances for future sustainable economic growth.
- Foster and support new sources of growth through innovation, environmentally friendly 'green growth' strategies and the development of emerging economies.
- Ensure that people of all ages can develop the skills to work productively and satisfyingly in the jobs of tomorrow.

OECD and sustainable development

The OECD significantly advances sustainable development through its comprehensive approach to economic, social, and environmental issues. Here's how the OECD contributes to sustainable development:

1. Economic growth and employment (SDG 8)
   - Inclusive growth: It promotes policies that foster inclusive economic growth, ensuring that the benefits of growth are shared broadly across society.
   - Employment and Skills: It supports initiatives to improve labour market outcomes, enhance skills and reduce unemployment, particularly among youth and marginalised groups.
2. Reducing inequality (SDG 10)
   - Income inequality: The OECD analyses income inequality trends and provides policy recommendations to reduce disparities and promote social inclusion.

  - Social policies: It promotes social protection policies and programs that support vulnerable populations and reduce inequality.
3. Education and lifelong learning (SDG 4)
  - Education systems: The OECD researches education systems and provides policy advice to improve education quality, equity, and efficiency.
  - Skills development: It emphasises the importance of lifelong learning and skills development in adapting to changing labour market demands.
4. Environmental sustainability (SDG 13, 14, 15)
  - Climate change mitigation: The OECD provides policy advice on reducing greenhouse gas emissions, promoting renewable energy, and enhancing climate resilience.
  - Biodiversity and ecosystems: It supports policies that protect biodiversity and manage natural resources sustainably.
  - Environmental performance reviews: It conducts environmental performance reviews to assess and improve member countries' environmental policies.
5. Sustainable infrastructure and innovation (SDG 9)
  - Infrastructure investment: The OECD promotes sustainable infrastructure investment that supports economic growth, reduces environmental impact and enhances quality of life.
  - Innovation policies: It supports innovation policies that drive sustainable development and address global challenges.
6. Good governance and institutions (SDG 16)
  - Public governance: The OECD guides on improving public governance, transparency and accountability, which are critical for sustainable development.
  - Anti-corruption: It combats corruption and promotes integrity in public and private sectors.
7. Global partnerships (SDG 17)
  - International cooperation: It facilitates international cooperation on economic, social and environmental issues, promoting policy coherence and mutual learning.
  - Development Assistance: It monitors development assistance, ensuring that aid is effective and aligned with sustainable development goals.

The OECD contributes to sustainable development by promoting inclusive economic growth, reducing inequality, enhancing education and skills development, supporting environmental sustainability, fostering sustainable infrastructure and innovation, strengthening governance and institutions and facilitating global partnerships. The OECD helps countries achieve the Sustainable Development

Goals and create a more equitable and sustainable world through research, policy advice, and international cooperation.

Asian Development Bank

The ADB was conceived in the early 1960s as a financial institution that would be Asian and foster economic growth and cooperation in one of the poorest regions in the world. A resolution was passed at the First Ministerial Conference on Asian Economic Cooperation held by the UN Economic Commission for Asia and the Far East in 1963, which set that vision on the way to becoming a reality. It opened on December 19, 1966, with thirty one members joining to serve a predominantly agricultural region. Takeshi Watanabe was ADB's first president and focused much of its assistance on food production and rural development.

Strategic Priorities of ADB

ADB's strategic priorities are based on three pillars given below given in Figure 23:

Priority pillar 1: Boost economic competitiveness to create more and better jobs.

Priority pillar 2: Inclusive provision of infrastructure networks and services.

Priority pillar 3: Address Climate Change mitigation and increase climate resilience.

Composition

ADB has 67 shareholding members, including 48 from the Asia and Pacific regions. The Board of Governors is the highest policymaking body, which comprises one representative from each member nation. The governors elect twelve members to form the Board of Directors, performing full-time duties at

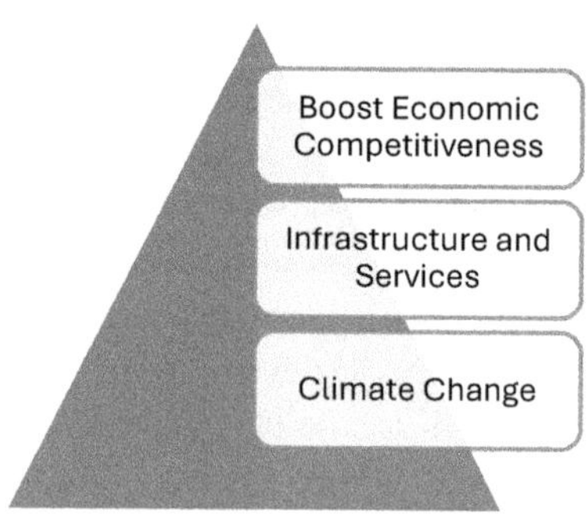

Figure 23 Priority pillars of ADB

the ADB Headquarters in Manila, Philippines. The directors supervise ADB's financial statements, approve its administrative budget, and review and approve all policy documents and all loan, equity, and technical assistance operations.

ADB's role in sustainable development

The ADB plays a significant role in advancing sustainable development in the Asia-Pacific region through its focus on inclusive and environmentally sustainable growth. Here's how the ADB contributes to sustainable development:

1. Promoting inclusive economic growth
   - Infrastructure development: The ADB finances infrastructure projects in transport, energy, water, and urban development to enhance connectivity, productivity and quality of life. Sustainable infrastructure is a crucial driver of economic growth and poverty reduction.
   - Private sector development: It supports private sector development by providing financing and technical assistance to small and medium-sized enterprises (SMEs), promoting entrepreneurship, and improving the business environment.
2. Reducing poverty and inequality
   - Social protection programs: It supports social protection programs that provide safety nets for vulnerable populations, including cash transfers, social insurance and labour market programs.
   - Inclusive finance: It promotes financial inclusion by improving access to financial services for underserved communities, including women, rural populations and low-income households.
3. Environmental sustainability
   - Climate change mitigation and adaptation: The ADB supports projects that reduce greenhouse gas emissions, promote renewable energy, and enhance climate resilience. This includes financing for energy-efficient technologies, sustainable transport and climate-smart agriculture.
   - Natural resource management: It funds projects that promote sustainable management of natural resources, including water, forests and fisheries, to ensure long-term environmental sustainability.
4. Improving health and education
   - Health programs: It invests in health infrastructure, services, and systems to improve healthcare access and outcomes. This includes projects on maternal and child health, disease prevention and strengthening of health systems.
   - Education initiatives: It supports projects that enhance access to quality education, improve learning outcomes, and promote lifelong learning.

This includes investments in schools, teacher training, and educational technologies.

5. Strengthening governance and institutions
   · Public sector management: It supports public sector reforms to enhance governance, transparency and accountability. This includes capacity building for public institutions and support for anti-corruption initiatives.
   · Decentralization and local governance: It promotes decentralization and local governance to empower local communities and improve service delivery at the regional level.
6. Enhancing regional cooperation and integration
   · Regional connectivity: It promotes regional cooperation and integration by financing cross-border infrastructure projects and facilitating trade and investment. This enhances economic growth and development across the region.
   · Regional public goods: It supports regional public goods such as transboundary water management, disaster risk reduction and regional health initiatives.
7. Specific initiatives and programs
   · Strategy 2030: The ADB's Strategy 2030 sets the course for the Bank's operations from 2019 to 2030, focusing on addressing poverty and inequality, accelerating progress in gender equality, tackling climate change, enhancing environmental sustainability and promoting regional cooperation.
   · Asian Development Fund (ADF): The ADF provides grants to the ADB's poorest member countries to help them achieve sustainable development goals. The ADF focuses on reducing poverty, fostering inclusive economic growth and building resilience to climate change and disasters.
   · Green bonds: The ADB issues green bonds to raise capital for projects with environmental benefits, such as renewable energy, energy efficiency and sustainable transport. This supports the transition to a low-carbon economy.
   · Pacific approach 2021-2025: This strategy outlines the ADB's support for sustainable development in Pacific island countries, focusing on building resilience, enhancing connectivity and promoting inclusive growth.

In summary, the ADB contributes to sustainable development by promoting inclusive economic growth, reducing poverty and inequality, enhancing environmental sustainability, improving health and education, strengthening governance and institutions and fostering regional cooperation. Through its financial assistance, technical support, policy dialogue, and knowledge sharing, the ADB helps member countries achieve their sustainable development goals and build a more prosperous, inclusive, and resilient Asia-Pacific region.[17]

Challenges faced by international agencies in sustainable development

International agencies such as the United Nations (UN), World Bank, International Monetary Fund (IMF), and various non-governmental organizations (NGOs) face several challenges in their efforts to promote sustainable development. These challenges can hinder their effectiveness and the realization of their goals. Here are some of the key challenges:

1. Implementation and compliance: One of the primary challenges is ensuring that international policies and frameworks are effectively implemented at the national and local levels. Although international agencies can provide comprehensive guidelines and objectives, their success largely depends on the willingness and capacity of individual countries to adopt and integrate these recommendations into their national policies. Political instability, corruption, lack of resources and competing national priorities often impede effective implementation and compliance.
2. Conditionalities and economic impact: The financial assistance provided by international agencies like the IMF and World Bank often comes with conditions, such as structural adjustment programs or austerity measures. These conditions can have adverse social and economic impacts, such as increased poverty and inequality, social unrest, and reduced public spending on essential services like health and education. While the intent is to promote fiscal responsibility and economic stability, the conditions can sometimes undermine sustainable development objectives by prioritizing short-term economic reforms over long-term social and environmental sustainability.
3. Capacity building and local adaptation: Building local capacity to manage development projects is crucial, but international agencies often face challenges in ensuring their capacity-building initiatives are effective and sustainable. Issues such as brain drain, where trained individuals migrate to more developed countries, and the mismatch between training programs and local needs can limit the impact of these efforts. Moreover,

17 Adapted from: Asian Development Bank, 'Sustainable development goals', <https://www.adb.org/what-we-do/topics/sdg>, accessed 4 August 2024.

the standardised approaches sometimes adopted by international agencies may not align with individual countries' unique contexts and challenges, necessitating more tailored and context-specific interventions.

4. Funding and resource allocation: While international agencies provide significant financial resources for sustainable development, the availability and allocation of funds can be inconsistent and subject to political and economic fluctuations. Donor fatigue, changing geopolitical priorities and economic downturns can lead to reduced funding for development projects. Additionally, securing financing can be bureaucratic and time-consuming, delaying the implementation of critical initiatives.
5. Coordination and fragmentation: The international development landscape is crowded with numerous agencies with mandates, priorities, and strategies. This can lead to fragmentation and lack of coordination among different actors, resulting in overlapping efforts, duplication of work and inefficient use of resources. Ensuring coherent and synergistic efforts among various international agencies, governments, and local organisations is essential but challenging.
6. Power dynamics and equity: The power dynamics between donor and recipient countries can create challenges in implementing sustainable development initiatives. Often, the priorities and perspectives of donor countries and agencies dominate, potentially side-lining the needs and preferences of recipient countries. Ensuring equitable partnerships where all stakeholders have a voice and can influence decisions is crucial for the success of development projects.
7. Monitoring and evaluation: Effective monitoring and evaluation of development projects are essential to ensure accountability, learn from successes and failures, and make necessary adjustments. However, international agencies often face challenges establishing robust monitoring and evaluation systems. Data collection can be hindered by a lack of capacity, political resistance and logistical difficulties. Additionally, the impact of development projects can be hard to measure, especially when dealing with long-term and complex issues like poverty reduction and environmental sustainability.
8. Environmental and social considerations: Balancing economic growth with environmental sustainability and social equity is a significant challenge. International agencies must navigate the complexities of promoting development that does not degrade the environment or exacerbate social inequalities. Projects funded by international agencies have sometimes been criticised for prioritizing economic development over environmental protection and social welfare, leading to unsustainable practices and negative impacts on local communities.

Conclusion

International agencies like the UN, WB, OECD, ADB and many others are working towards achieving the SDGs. International agencies have had mixed success in pursuing sustainable development, reflecting significant achievements and ongoing challenges. On the positive side, organizations like the World Bank, the Asian Development Bank (ADB), and the United Nations Development Programme (UNDP) have substantially contributed to poverty reduction, infrastructure development, and promoting inclusive economic growth. Initiatives such as the World Health Organization's (WHO) efforts in combating global health crises and the Food and Agriculture Organization's (FAO) work in promoting food security have also demonstrated these agencies' critical role in fostering sustainable development. The United Nations International Children's Emergency Fund (UNICEF) has also improved child health, education, and protection, contributing to long-term sustainable outcomes. However, despite these successes, significant challenges remain.

The International Monetary Fund (IMF) has been criticised for policies that sometimes exacerbate economic inequalities and social tensions in developing countries. Furthermore, the effectiveness of international agencies is often hampered by bureaucratic inefficiencies, lack of coordination, and insufficient funding. The complexities of climate change, persistent poverty, and geopolitical conflicts continue to pose substantial obstacles to sustainable development, indicating that while international agencies have achieved notable successes, there is still a long way to go to realise sustainable development goals. While international agencies play a vital role in promoting sustainable development, they face numerous challenges that can impede their effectiveness. Addressing these challenges requires a multifaceted approach, including greater emphasis on local adaptation, enhanced coordination among agencies, more equitable partnerships, and robust monitoring and evaluation systems. International agencies rank all member nations on various indices like human development, world happiness, democracy, and other parameters. They help examine the countries' policies with the best scores to make them the best practices to be followed.

Citizen-centric Governance

Introduction

Government administration refers to the management of government activities, while development administration's primary focus is the identification of the beneficiaries of developmental efforts. Developmental initiatives and actions should improve people's lives and not merely be an exercise in goal achievement. Aligning development activities with the real needs and interests of the most vulnerable or marginalized is the most important way to demonstrate sustainability in development. The ultimate beneficiary of every form of development, whether social, physical, economic or cultural, is the citizen. Since all development is carried out in the name of the nation's citizens, it must be focused on their well-being. In a democratic society, governments, public institutions and systems must operate with the citizens' interests at heart. Democracy is defined as government by the people, of the people and for the people. Therefore, democratic governments that by definition consist of elected representatives hold the responsibility to serve the public, including in development contexts. The core principle behind all policy-making at different government levels should be to enhance the quality of life for its citizens.

Today, the United Nations' seventeen Sustainable Development Goals (SDGs) of the 2030 Agenda, introduced on January 1, 2016, represent a framework for citizen-focused governance. It is the responsibility of the government to ensure that its efforts prioritise the welfare of citizens, addressing their desires and requirements, while involving them in democratic processes through decentralisation and delegation. The 73rd Constitutional Amendment in India marked progress toward promoting citizen-centered governance. At the heart of good governance is citizen involvement. The success of public policies depends heavily on their effective implementation. Without proper governance, policies may fail, depriving citizens of the benefits they were designed to offer.

Citizen-centric governance or citizen-centric development administration must be based on the principle of 'minimum government, maximum governance'. It implies that there must be 'responsive governance' in favour of the needs of the citizens and directing its systems and institutions towards equitable distribution of benefits. It works with the people by 'citizen participation' via various CSOs. It manifests direct democracy using multiple tools and mechanisms like the Citizens' Charter (Freedom to Information Act or any other name it is known by), regulatory bodies, social audits, and the like. The bottom-up approach to developing citizen-centric governance, in which the citizen participates in the development process, involves increasing decentralisation of power away from the union government and closer to the grassroots levels via local governance. Citizen-centric development administration focuses on not only the benefits to be derived by the citizens but also their part in governance to lead to improved outcomes such as better service delivery and redressal of grievances with absolute accountability. The focus of all citizen-centric governance for development is to aim for the achievement of sustainable development goals.

All modern democratic societies are moving towards citizen-centric governance, which is more accessible for governments, policymakers, front-line staff, and ministers because they know who the policy targets and what the citizens require. Citizen-centric governance is adopted by simplifying processes and procedures, making life easier for all.

Building sustainable futures through citizen-centric governance

Citizens are the predominant point of achieving sustainable development through citizen-centric governance. The decisions taken in this regard have to align people's needs and the policies. This approach has adopted the principles of transparency, accountability and inclusivity so that people can be a part of the future. Building robust, fair and sustainable communities is one of the sustainable development goals established by the United Nations. Governments are the primary agency who can contribute to sustainable futures by promoting a collective action programme. The major thrust is on enabling the common people with powers to influence development programmes and policies. The governments have the capability to understand the current global socioeconomic challenges which helps them to explain them to the people. Citizens can be more responsive and participative only if they are aware and updated about all the aspects affecting them. Citizen-centric governance is a revolutionary approach that has reconceived the bond between the governments and the governed. The model keeps the citizens at the frontline of sustainable development by engaging with them. The more the governments meet the expectations of the people, the more the latter will trust the former. The people will accept the government policies more readily.

Features of citizen-centric governance

Conventional public administration adhered to rigid systems, institutions, and processes, leaving little room for flexibility. Once policies and plans were

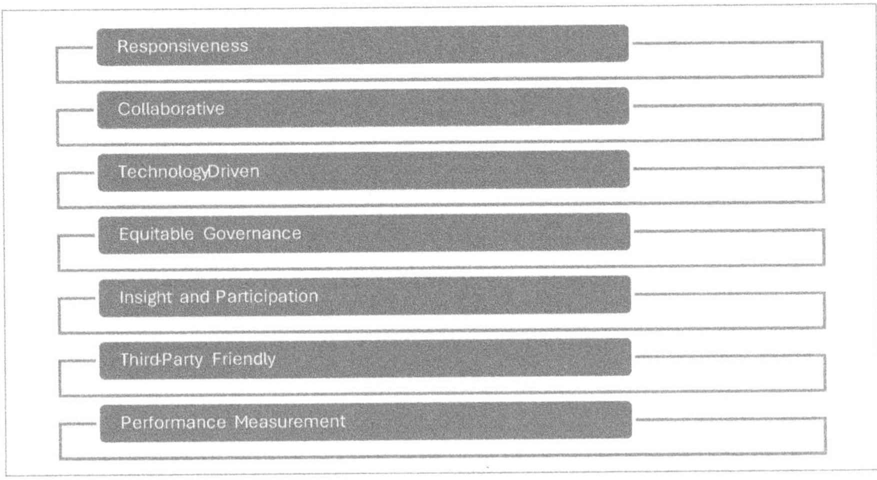

Figure 24 Features of citizen-centric governance

developed, they had to be implemented strictly, often neglecting risk assessments or unforeseen impacts. In contrast, modern public administration has adapted its systems to meet the evolving developmental needs of society. The key elements of citizen-centric governance, as shown in Figure 24, are outlined and explained below.

1. Responsiveness: Responsiveness is one of the World Bank's worldwide good governance indicators which is the main goal of citizen-centric governance. Rule of law is essential in ascertaining that people's needs are fulfilled without resorting to populist policies. Along with responsiveness, government agility is crucial. For instance, when launching any project, the initial plan often needs adaptation, requiring flexibility to meet public needs. India's current Prime Minister, Narendra Modi, introduced the concept of 'SMART' development administration, which stands for Simple, Moral, Accountable, Responsible, and Transparent, capturing the core aspects of citizen-centric governance.
2. Collaborative: The ways of conventional public administration has the government departments working in silos without any channels of information sharing among them. Collaboration is a necessary component to improve coordination among the departments to attain successful outcomes. On the other hand, good governance in the modern world gives the following key points:
   · Strong and stable political leadership endorsed with absolute majority
   · Out-of-the box innovative approaches

· a collaborative perception of desired results[1]

3. Technology-driven: The introduction of new technology is never easy especially in the public sector. E-governance and digitisation are two pillars of citizen-centric governance that help in the achievement of transparency and accountability. Citizens are accredited with power to access information via digitised service delivery regarding the status of services that they have been promised. Programmes like e-Seva in India help in processes like fee collection and service delivery. The Government of India started online service integration through portals like 'One-Stop Shops' as a part of 'Whole of Government' approach. Thus, the technology-driven governance, a characteristic of New Public Management, with the help of tools like ICT and Internet of Things, promotes citizen-centric governance.[2]
4. Equitable governance: Ensuring that the government treats all citizens equally and focuses on uplifting weaker sections of society has long been a cornerstone of citizen-focused governance. Equitable growth is essential to prevent widening socio-economic divisions. The concept of equitable governance gained prominence in the 1960s when public administration scholars like George Fredrickson emphasised that delivering services such as healthcare, education, welfare and law and order efficiently and equitably is crucial. If public administration fails to ensure social equity, it becomes part of the problem. This idea was discussed during the Minnowbrook Conferences.
5. Insight and participation: As discussed earlier, citizen participation is an important element of citizen-centric governance but it necessitates government insight. Rational decision-making concerning the needs of the citizens can be enhanced by using decentralization and delegation. Feedback information can help in improving policy formulation and execution by the governments.
6. Third-party friendly: Governments cannot provide public services solely on their own due to the multiplicity of demands on them. They must rely upon third party providers but for that they must have a friendly and good network.
7. Performance measurement: Evaluating service delivery is crucial in citizen-focused governance as it helps governments understand the extent of citizen satisfaction. Implementing benchmarks and key performance indicators

---

1 Peter Wilkins, John Phillimore, and David Gilchrist, *Working Together: Evidence on Collaboration from the Reports of Independent Watchdogs* (Melbourne: ANZSOG, 2015).

2 Ravinder Kumar Bagga, *The State, IT and Development* (New Delhi: SAGE Publications, 2006).

(KPIs), establishing reward systems for high performers, and developing training programs are essential for fostering knowledge-based organizations and enhancing operational management.[3]

Citizen-focused governance is assessed through worldwide indicators for governance, though scholars like Ivanya and Shah have highlighted certain limitations and proposed frameworks for better comparative assessment of governance quality across countries and time. These indicators capture citizens' perceptions of governance in their respective countries.

Emergence of citizen-centric governance: The core of good governance

Evidence from ancient civilizations like the Sumerians, Babylonians, and the Indus Valley shows that rulers planned cities with proper infrastructure to meet the needs of citizens. Historical texts also reference rulers addressing citizen grievances in public forums. The Indian treatise Arthashastra, written by Kautilya in 320 BC, stressed the importance of the welfare of the populace. Kautilya famously said, 'In the happiness of his subjects lies the happiness of the ruler; in their welfare lies his welfare'.

After a two-century gap during which the world was focused on economic recovery post-war, public administration rediscovered citizen-centric principles in the 1960s and 1970s through New Public Administration and New Public Management. In 2008, the World Bank introduced Worldwide Indicators for Good Governance, marking the beginning of formalised citizen-centric governance. The period for Millennium Development Goals ended in 2015 and were replaced by Sustainable Development Goals. Their primary objective is to promote and improve citizen-centric governance to attain equitable development and poverty alleviation.

Principles of citizen-centric governance

Given below are some core principles recommended by the Second Administrative Reforms Commission, India, for citizen-centric governance:

1. Rule of law: The administrative agencies must have a zero-tolerance strategy towards crime to establish compliance and public order. The zero-tolerance

3 World Bank, *Citizen-Centric Governance Indicators: Final Report* (Washington, DC: World Bank, 2018), <https://documents1.worldbank.org/curated/en/775701527003544796/pdf/126399-WP-PUBLIC-CitizenCentricGovernanceIndicatorsFinalReport.pdf>, accessed 25 August 2024.

strategy can be put in place to enable better law compliance. It must be supported by establishing statistical databases and artificial intelligence technology that can track any corruption among the public officials. The community can be engaged in deterrence of crime.

2. Vibrant, responsive and accountable institutions: The institution that have been vested with the powers and have appropriated funds can help in effective implementation with trained officials. When governance is citizen-centric, transparency and openness become driving forces behind institutional operations.
3. Active participation of citizens: Active engagement and involvement are crucial for effective governance and can be significantly enhanced through decentralising delivery mechanisms. Citizens and stakeholders are more directly involved in decision-making processes by distributing authority and responsibilities to lower levels of government, such as the Panchayati Raj Institutions in India. This multi-tiered approach brings governance closer to the people and ensures that local needs and priorities are better understood and addressed. Decentralization fosters a sense of ownership among citizens, encouraging them to actively contribute to the planning, implementing, and monitoring of policies and services, ultimately leading to more responsive and accountable governance.
4. Transparency: The decision-making processes can be open and accountable when citizen-centric governance follows the principle of transparency. Confidence building measures are vital to the maintenance of standards of public service delivery. If the institutions declare their day-to-day administrative operations on their websites and manuals, people will understand if they are being given prompt and correct services or not. Public officials can be held accountable if the service is delayed or wrong. The legitimacy of public administration is augmented with active citizen engagement as a result of transparency in the administrative processes and procedures of an institution.[4]
5. Civil service reforms: Civil servants manage government institutions. Citizens' experience is based on how civil servants resolve their issues, whether they are courteous and empathetic or have a snobbish attitude. Civil services should be designed to prioritise 'citizen-centric governance', ensuring that the entire governance structure serves the people effectively. The civil service must become more responsive, transparent and

---

4 *Center for Good Governance, Ushering in Transparency* (Hyderabad: Center for Good Governance, 2017), <https://cgg.gov.in/core/uploads/2017/07/Ushering_in_Transparency.pdf>, accessed 25 August 2024.

accountable to the needs of citizens. Public servants should have the necessary tools, training, and ethical grounding to focus on the public good rather than bureaucratic inertia. Emphasizing service delivery, citizen participation and inclusivity transforms the governance apparatus into a dynamic entity that listens to and acts on the public's concerns, fostering trust and strengthening democratic governance.

6. Ethical governance: Ethics guides governments, institutions and people equally to behave per standards that a society enforces on itself. Corruption and misuse of office are direct consequences of lapses in compliance with ethics. A lack in ethical judgement on the part of citizens influences their behaviour also making them liable to use bribery as a tool to get work done. Thus, ethical governance is a foundational principle of citizen-centric governance.
7. Police reforms: Police reforms extend beyond merely enacting laws, formulating policies and setting up institutional structures. Implementing processes that ensure these reforms serve citizens effectively is essential. Citizen charters with clear service standards and penalties for non-compliance play a crucial role in this effort. Establishing independent and empowered anti-corruption agencies is also vital in maintaining integrity within the police force. Involving citizens through innovative tools and mechanisms helps enhance transparency and accountability, leading to a more responsive and citizen-focused approach to policing.
8. Evaluation of quality of governance: Evaluating the quality of governance is essential to ensure that policies and programs achieve their intended outcomes. The adage 'what is not monitored, never gets done' highlights the importance of regular assessments. Evaluation can be conducted internally, where government bodies themselves review and monitor their processes and effectiveness. Alternatively, independent external agencies can provide an unbiased perspective, offering critical insights into areas needing improvement. These evaluations help identify inefficiencies, promote accountability, and ensure that governance remains transparent and responsive to citizens' needs. Regular assessments ultimately contribute to a more effective and citizen-centric governance framework.
9. Principle of subsidiarity: The Second Administrative Reforms Commission in India recommended the implementation of the subsidiarity principle in local governance.[5] This principle promotes the following three values:
10. Ownership and engagement: Decision-making at the local level boosts efficiency, self-reliance and innovation. People feel more involved in

5 'Subsidiarity', as defined in various national constitutions.

developmental programs and socioeconomic initiatives when they are a part of the decision-making process. They get a sense of ownership towards the progressive programmes that are being implemented for them.

11. Subsidiarity as the foundation of democracy: The principle of subsidiarity aligns with the core assumptions of democracy, which are:
    - All citizens possess equal standing.
    - The citizen is the ultimate authority.
    - Citizens have the ability to determine what serves their best interest.
    - These three assumptions must be implemented to establish the complete legitimacy of a democratic system.
12. Empowerment: Citizen-centric governance helps empower vulnerable groups by enabling them to influence public policy and improve service delivery. It aims to bridge the gap between marginalised individuals and the stakeholders who create and implement policies that affect their lives. In essence, citizen-centric governance places a strong emphasis on addressing the needs and concerns of individuals, ensuring transparency, accountability, and responsiveness within public administration. Embracing inclusivity, participation, and service orientation enhances trust, improves service delivery, and fosters a more equitable and effective societal framework.

Citizen's participation

The ultimate beneficiary of citizen-centric governance is the citizen of all policies and plans of a government. Citizen participation in the mechanisms, processes and the agencies through which decisions are made and executed is a foundational tenet of good governance. The UNDP (1997) states that citizens' groups and communities safeguard their ideas, express their interests, exercise their legal rights, fulfil their duties, and mediate differences. A government's perspective towards the citizens has evolved from perceiving them as passive beneficiaries or customers to active members in the developmental work. Over the past five decades, there has been a conceptual change in development administration from considering citizens as passive recipients to active participants. The change in development administration has been primarily from a 'top-down' approach to a 'bottom-up' approach. This follows from the principle of subsidiarity discussed earlier. The power is decentralised from the national level to the local authorities, enabling the grassroot people.

Citizens' participation is paramount to citizen-centric governance in having some sway over making decisions and regulating resources that will alter their lives. From an ideological perspective, involving citizens directly in governance bolsters democracy by improving and supplementing traditional representative democracy, making it more responsive and participatory at the grassroots level.

Citizen participation and engagement in sustainable development governance are essential for crafting inclusive, responsive and effective policies that tackle the complex challenges of sustainability. By involving citizens in decision-making, diverse perspectives are considered, resulting in more equitable and locally relevant solutions. This participatory approach empowers individuals and communities, developing ownership and responsibility to achieve sustainable targets. Engagement can take various forms, such as public consultations, community forums, participatory budgeting and collaborative governance models. By actively involving citizens in planning and implementing policies, governments can tap into local knowledge, creativity, and innovation. This not only improves the quality of decisions but also enhances their legitimacy and public acceptance. When citizens feel their voices are acknowledged and valued, they are more likely to support and actively participate in sustainable development efforts.

Citizens are no longer viewed merely as recipients of technical expertise or beneficiaries of externally directed development programs. Instead, they are recognised as equal partners in the development process. Promoting public participation as a democratic right is now seen as a core objective in all development initiatives, rather than just a means to development. Engaging citizens in governance is a collaborative agreement between government bodies and the public, encouraging maximum involvement. This fosters better service delivery, influences public policy, ensures effective grievance redressal, and leads to enhanced outcomes.

Citizen's participation and good governance

Citizen participation is the key to sustainable development as the citizens' needs are lined up with sustainable development goals. Active citizen engagement helps the officials understand their socioeconomic cand environmental issues and they work towards specific solutions. This inclusiveness ensures that development projects are practical and equitable, addressing the needs of all segments of society, including the marginalised. Active citizen engagement develops a commitment to results and taking the lead among community members, encouraging them to support and sustain these initiatives over the long term. Transparency and accountability are boosted, thereby, reducing the risk of corruption and misuse of resources. Citizen participation promotes better governance practices by creating a more informed and involved populace, which is essential for achieving sustainable outcomes that benefit current and future generations.

Good governance can be improved with citizen participation as outlined here:

1. Accountability: Increasing awareness among citizens regarding their rights and government sustainable development processes empowers them to demand for accountability. It empowers citizens to call for accountability, enhancing the government's responsiveness, efficiency, and effectiveness. Citizen participation promotes transparency and accountability in governance. It encourages governments to be more open about their decision-making processes and to justify their actions to the public. This helps reduce corruption and the misuse of resources, ensuring that development initiatives are truly in line with the needs of the population.
2. Sustainability: It helps to make government programmes and services more effective and sustainable. Citizen engagement is vital for the sustainability of government programs as it fosters ownership, accountability, and effectiveness. Citizens develop a personal stake in the sustainable development work that helps in long-term support. Communities can address their unique environmental, social, and economic challenges by actively involving citizens in decision-making.
3. Helping the marginalised: Quality of life of citizens is improved by enabling the marginalised sections of the society as public policy and service delivery is upgraded. Voice is one of the indicators of Worldwide Governance Indicators given by the World Bank which can be heard only if there is citizen inputs. The government must make an effort to convince low-income individuals in participating in the decision-making process. The latter's needs can be included in the policymaking process itself so that the people get development that is the most essential.
4. Grassroots democracy: Citizen participation reinforces grassroots democracy when the lowest socioeconomic levels of the society are empowered. Mahatma Gandhi suggested to the policymakers that said that they must ask themselves if the poorest and the weakest man in the society is going to be benefitted with the policy or not. Once that is achieved, grassroots democracy can be strengthened.

Methodology for citizen participation

The Second Administrative Reforms Commission of India recommended some critical methods to aid in citizen participation in governance:

1. Citizens seeking information: Access to information is vital for enabling citizen participation. Providing comprehensive details (on procedures, prices, application forms, grievance contacts, etc.) is the first step in assisting citizens to take part in the government effectively.

2. Citizens giving suggestions: Sustainable development necessitates sustainable participation of citizens in government efforts. Public hearing can be conducted in town halls or community halls to listen what people suggest regarding their issues. Surveys can be carried out after the developmental activity is completed to gather feedback from the beneficiaries to make any further modifications if needed. These suggestions, when combined with governmental expertise, can result in better policymaking and implementation. Simple initiatives like suggestion boxes, online portals and regular consultations with citizen groups can facilitate this process.
3. Collaborative evaluation: Ensuring better public service delivery is the aim of citizen participation, thus, a grievance redressal system can be established so that there is a collaborative effort in evaluating the services. The government employees will remain more efficient in their work so that their feedback is not negative. ICT can enhance both the monitoring and evaluation systems to ensure better service delivery for example in the case of Citizens' Charters.

   Public accountability by citizens: Government agencies can utilise mechanisms like citizen report cards and social audits to assess the quality of their service delivery. Efficiency, effectiveness, timeliness and customer satisfaction can be some of the indicators to assess the public services.
4. Active citizen participation in administration and decision-making: According to the Second ARC's report on local governance, robust citizen participation gives the public continuous access to decision-making beyond periodic consultations. Citizens' groups act as key stakeholders with whom the government must coordinate. Participatory processes like municipal budgeting and referendums allow citizens to directly vote on proposals related to public policies, projects, and laws.
5. Social audit: A social audit involves involving all parties engaged in public services to appraise the development activities especially with reference to sustainability. It promotes government transparency, encourages social engagement, and ensures accountability by assessing public agencies' performance from a societal perspective.
6. Enabling women's participation: It is more difficult for women to approach and access information or some service from a government office due to sociocultural mores and seclusion norms. So, women must be included in all efforts to encourage citizens' participation in the earlier mentioned points. Citizen-centric governance can be strengthened with women participation in the development process.
7. The physically challenged: Many international agencies like the World Bank and the United Nations Development Programme use an inclusive

approach for addressing the needs of the physically challenged. They forma large part of our society and cannot be left deprived of the progress and sustainable development initiatives. Public officials and all other stakeholders involved in the activity must ensure that there is appropriate and accessible infrastructure for them. Monitoring and evaluation mechanisms ensure continuous improvement, while partnerships with NGOs and the private sector create a supportive ecosystem. These strategies cultivate an ecosystem where all citizens, including those with physical disabilities, can fully participate in society and access public services effectively.

8. E-governance: Citizen engagement mechanisms: The following four methods can be used to promote citizen participation in e-governance projects:
   - Information sharing: Spreading a wide range of information raises awareness among citizens, elected representatives and other stakeholders. This can be done through citizens' charters, regular meetings, mobile-based voice and text services, and web platforms to share key information.
   - Consultation: Regular consultative meetings should be conducted with stakeholders, including citizens, elected representatives, local civil society groups and others. These consultations can happen online or in-person and at different stages of the project cycle.
   - Joint assessment: Citizen engagement can be strengthened through participatory assessments and monitoring with stakeholders, especially service seekers. Collaborative assessment can be done through joint citizen monitoring or representative meetings.
   - Collaborative decision-making: Using participatory planning techniques, citizens and stakeholders can be involved in shared decision-making processes, fostering collaboration on governance issues.

Hence, the given points of citizens' participation improve service delivery. There can be no single mechanism to achieve improved citizen participation, but a comprehensive approach can be followed:

- reviewing policies and practices in every department or public agency
- modifying administrative procedures as required
- assigning a senior-level officer the responsibility of institutionalizing citizens' participation in governance
- conducting performance management reviews to ensure the effectiveness of citizen participation initiatives

Mechanisms for citizen-centric governance

Globally, there have been various initiatives to achieve citizen-centric governance, like using technology, the Right to Freedom, the Consumer Protection Act and the Citizen's Charter. They are discussed further.

Citizen's charters

A citizen's charter is a formal document that outlines the rights and responsibilities of citizens concerning public services provided by government agencies. It serves as a commitment from the government or public organization to deliver specific service standards, such as timelines, quality, and accessibility. The charter typically includes information on the services offered, the expected service standards, grievance redress mechanisms, and contact details for further assistance. The primary purpose of a citizen's charter is to enhance transparency, accountability, and responsiveness in public service delivery, empowering citizens to know what to expect and how to seek redress if services are not up to standard.

Advantages of citizen's charter: The advantages of citizen's charter are as follows:

- Aligning public service with citizens' expectations: Citizen charters serve as an interface between the public's expectations and the actual service delivery by the government. Often, there's a gap between the two, resulting in citizen dissatisfaction with the quality of services. A charter helps bridge this gap and can lead to better planning and policy formulation, avoiding misunderstandings regarding citizen needs. Effective strategies must be established at various administrative and political levels, focusing on clarity, interaction, and monitoring.
- Improving performance: While governments often have sound plans and projects in place for public welfare, implementation and quality become problematic. Citizen charters can address this issue by improving the government's performance and the quality of services provided.
- Ensuring time-bound services: Citizen charters guarantee that services are provided within a specified timeframe. If services are delayed, the value is diminished. If legislated, this can ensure timely delivery, and in some cases, penalties may be imposed for failing to meet deadlines.
- Enhancing transparency: A key goal of citizen charters is to enhance transparency, enabling citizens to access necessary services without corruption. Simplified processes and procedures play a critical role in achieving this.

Principles of citizen charters:

1. Transparency: Charters must strive to enhance transparency by fostering openness and abandoning secrecy in service delivery procedures. Efficient contact points and facilitation counters must be included.
2. Quality: Improving the quality of public services is a primary goal of citizen charters. Pre-defined service standards help ensure accountability. Tools like Total Quality Management (TQM) can help assess and improve service quality based on agency-specific data.
3. Choice: Citizen charters promote customer choice, enabling individuals to select services that meet their preferences. However, customer choices vary based on their economic status, and they need to be aware of the latest developments to make informed decisions.
4. Accountability: The public sector must be accountable to citizens. Citizen charters help assign responsibility to officials for service delivery failures. Accountability is maintained through monitoring, evaluation and grievance redressal.
5. Standards: Service standards should be clearly stated in citizen charters. Originally, the UK government emphasised setting service delivery standards, and maintaining them is essential for success. Key principles include:
   - Prioritizing the customer
   - System analysis
   - Continuous improvement efforts
   - Stressing facts and figures
   - Building a strong client-supplier relationship
6. Value: Public organizations must ensure they deliver value for taxpayer money by using resources efficiently. This principle ensures that citizens receive quality services in return for their contributions.

These principles should guide the formulation of citizen charters to empower citizens and enhance service delivery.

The Consumer Protection Act

The Consumer Protection Act is a vital piece of legislation designed to safeguard the rights and interests of consumers. Enacted to promote fair trade practices and ensure that consumers are not exploited or subjected to unfair practices, the Act provides a framework for addressing grievances related to defective goods, deficient services and unfair trade practices. It establishes consumer courts at various levels where consumers can file complaints and seek redressal. The Act also mandates the establishment of the Central Consumer Protection Authority

(CCPA) to regulate matters related to consumer rights, such as false advertising and unsafe products. Additionally, the Act empowers consumers with the right to information, choose, and be heard, fostering an environment of trust and fairness in the marketplace. The Consumer Protection Act is crucial in maintaining a balanced and equitable consumer market by holding businesses accountable.

The Consumer Protection Act, or its equivalent, is present in many countries, though the specific name, provisions, and enforcement mechanisms can vary widely. Most nations have established laws to safeguard consumer rights, protecting against unfair trade practices, defective products and substandard services. These laws typically allow consumers to seek redressal through designated consumer courts, ombudsmen, or regulatory bodies. Some example:

- United States: The Federal Trade Commission (FTC) enforces consumer protection laws such as the Fair Credit Reporting Act and the Consumer Product Safety Act.
- United Kingdom: The Consumer Rights Act 2015 covers issues like goods, services and digital content, ensuring consumers can seek redressal for faulty products.
- Australia: The Australian Consumer Law (ACL) protects consumers from misleading conduct and ensures rights to purchase goods and services.
- India: The Consumer Protection Act 2019 is a comprehensive law that protects consumer interests in India.[6] While consumer protection principles are universal, the legislation's specifics and enforcement methods differ by country.

For example, suppose a consumer buys a smartphone that has a faulty battery, and the company refuses to replace or repair it within the warranty period. In that case, the consumer can approach the district consumer forum to seek redressal. The court may order the company to replace the defective product, provide a refund, or pay compensation for the inconvenience caused. This process ensures that consumers are protected from unethical business practices and have a legal avenue to assert their rights.

Digitalised citizen-centric governance

After e-governance became the buzzword in the last two decades, it became essential to have some guidelines about digital governance to bring governments

6 United Nations Conference on Trade and Development, *Consumer Protection in Electronic Commerce* (Geneva: UNCTAD, 2016), <https://unctad.org/system/files/official-document/ditccplpmisc2016d1_en.pdf>, accessed 25 August 2024.

and citizens closer to each other. All that citizen-centric governance has accepted can be delivered with the help of new technologies like cloud computing, social media, and mobile technology, which integrate into the day-to-day lives of citizens. The OECD presented a document based on the recommendations on digital government strategies developed by the Public Governance Committee (PGC).

The OECD Digital Governance Index (DGI)

The Digital Government Index (DGI) evaluates how countries integrate technology and data to enable a human-centric public sector transformation. The OECD has outlined essential foundations necessary for this, as detailed in the OECD Recommendation on Digital Government Strategies. However, it's crucial to note that the DGI does not assess specific digital government processes or service usage by citizens – other international benchmarks cover those. Instead, it measures six dimensions of digital government maturity:

- Digital by Design: Evaluates the coherence of digital government policies and digital tools in public sector activities.
- Data-Driven: Assesses advancements in data governance and the ability to access, share, and reuse data across the public sector.
- Government as a Platform: Looks at using common resources like digital identities and software to standardise government processes.
- Open by Default: Measures the transparency and efforts to engage stakeholders using technology and data.
- User-Driven: Focuses on how effectively governments incorporate user needs into public service delivery.
- Proactiveness: Evaluates the ability to anticipate and meet user needs proactively.[7]

The Digital Government Index (DGI) evaluates digital government maturity across six key dimensions, as illustrated in Figure 25:

- Dimension 1 - Digital by Design: Evaluates how digital government policies empower the public sector to coherently utilise digital tools and data in policy formulation and public service transformation.

7 Organisation for Economic Co-operation and Development, *The OECD Digital Government Policy Framework* (Paris: OECD, 2020), <https://www.oecd.org/en/publications/2023-oecd-digital-government-index_1a89ed5e-en.html>, accessed 25 August 2024.

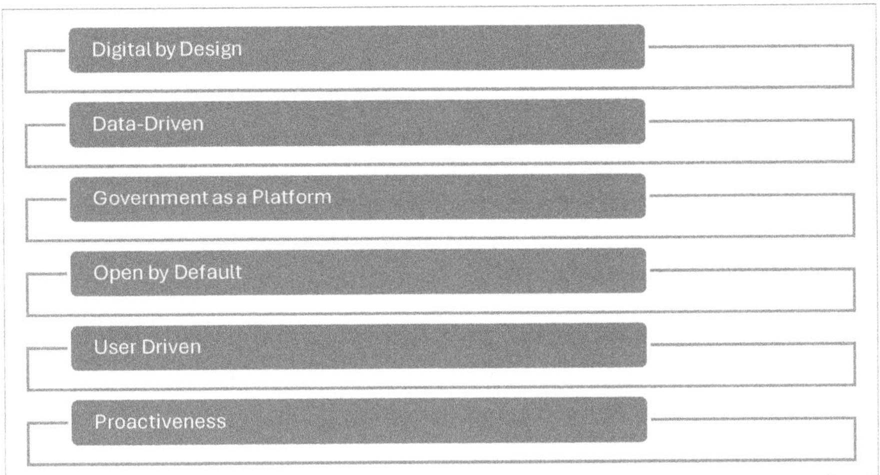

Figure 25 The six dimensions of the OECD Digital Government Index. Source: Organisation for Economic Co-operation and Development, The OECD Digital Government Policy Framework (Paris: OECD, 2020), <https://www.oecd.org/en/publications/2023-oecd-digital-government-index_1a89ed5e-en.html>

- Dimension 2 - Data-Driven: Assesses the government's progress in establishing governance structures and enablers for data access, sharing, and reuse across the public sector.
- Dimension 3 - Government as a Platform: Measures the deployment of standardised building blocks, such as guidelines, tools, digital identity, data and software, to support the transformation of government processes and services across the public sector.
- Dimension 4 - Open by Default: Looks beyond releasing open data, assessing efforts to encourage the use of technologies and data to engage and communicate with various stakeholders.
- Dimension 5 - User-Driven: Evaluates the government's ability to prioritise user needs in the design and delivery of public policies and services.
- Dimension 6 - Proactiveness: Measures the government's ability to anticipate the needs of users and service providers to proactively deliver government services.

Smart governance

Smart governance involves using technology, data and innovative practices to enhance governance processes by improving efficiency, effectiveness, transparency and inclusiveness. It involves leveraging digital tools, data analytics and ICT (Information and Communication Technology) to facilitate better decision-making, enhance citizen engagement, and streamline public services.

The World Bank defines smart governance as a framework that leverages digital technologies and data-driven approaches to enhance the effectiveness, efficiency, transparency and inclusiveness of governance processes. It focuses on using innovative ICT (Information and Communication Technology) to improve the way governments interact with citizens, deliver public services, and make decisions. Smart governance seeks to create more responsive and accountable institutions by integrating technology into various aspects of governance, from policy-making to service delivery.

Key elements of smart governance according to the World Bank

- Citizen-centric services: Smart governance prioritises the needs and preferences of citizens in designing and delivering public services. This includes simplifying interactions with the government, reducing bureaucratic burdens and enhancing user experience.
- Data-driven decision making: Governments use data analytics and big data to inform policy decisions, predict trends, and respond to citizen needs more effectively. This data-driven approach helps create evidence-based policies that are more likely to achieve desired outcomes.
- Transparency and accountability: By utilizing technology, smart governance increases transparency in government operations, making it easier for citizens to access information and hold officials accountable. Open data initiatives and digital reporting are examples of this.
- Interconnectivity and interoperability: Smart governance emphasises the importance of interconnected systems within the government, enabling seamless data exchange and collaboration across various departments and agencies.
- Citizen engagement: The World Bank emphasises the role of technology in fostering greater citizen participation in governance. This includes the use of e-participation tools, online platforms for feedback and digital forums for public consultations.
- Sustainability and inclusiveness: Smart governance also considers the sustainability of its initiatives and seeks to include all segments of society, particularly vulnerable groups, ensuring that no one is left behind in the digital transformation.

In essence, the World Bank views smart governance as an approach that uses technology not just for digitalization but as a means to fundamentally transform how governance operates, making it more effective, transparent, inclusive, and responsive to the needs of citizens.Some examples of Smart Governance are smart cities in urban areas to manage resources like water, electricity, and transportation more efficiently or E-Government portals where citizens can access

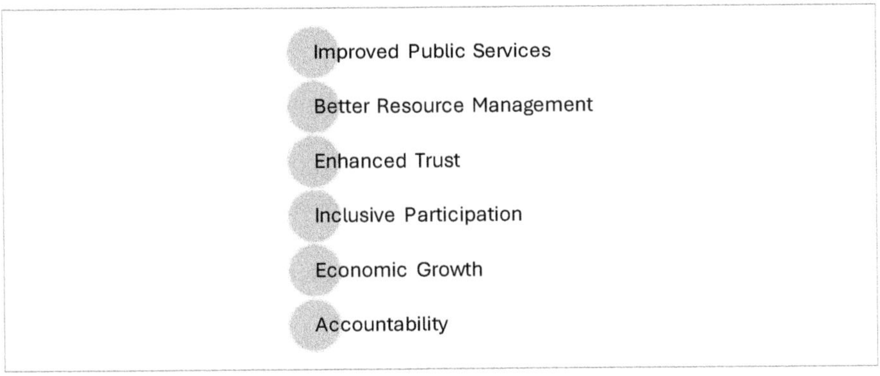

Figure 26 Benefits of smart governance

government services, submit applications, and track the status of their requests. Open Data Initiatives include governments making their data available to the public to increase transparency and allow for the development of innovative solutions by citizens or businesses.

Smart governance is beneficial in many ways but the important ones are very briefly given below in Figure 26.

- Improved public services: faster, more reliable and user-friendly public services
- Better resource management: efficient use of resources through data-driven insights
- Enhanced trust: increased transparency and accountability build greater trust between citizens and the government
- Inclusive participation: broader citizen engagement through digital platforms
- Economic growth: efficient governance can attract investment and foster innovation.
- Accountability: data-driven executive decisions will promote transparency and accountability

Smart governance and digitalised governance are related concepts, but they are not the same. While both involve the use of technology in governance, smart governance is a broader and more holistic concept that encompasses digitalised governance as one of its components. Digitalised governance primarily focuses on the transition of government services and processes from analogue or manual systems to digital formats. This involves the use of ICT (Information and Communication Technology) to improve the efficiency and accessibility of government operations and services. Some key elements of digitalised governance are E-Government services, such as online tax payments, digital applications for permits, digital records management and many other areas. The main goal is to make government services more efficient, accessible, and user-friendly through digital means.

Smart governance, on the other hand, takes digitalisation a step further by integrating technology, data and innovative practices into governance to enhance decision-making, citizen engagement and sustainability. It is more strategic and aims to create a connected, data-driven and participatory governance model. The goal of smart governance is to create a more responsive, transparent, efficient, and inclusive governance model that leverages technology and data to address complex challenges and improve public outcomes.

Both concepts are interrelated. Digitalised governance is often a foundational element of smart governance. Without digitalization, the advanced features of smart governance, such as data-driven decision-making and interconnected systems, would not be possible. Smart governance builds on digitalised governance by adding layers of data analysis, citizen engagement, and strategic planning to create a more comprehensive approach to governance. While digitalised governance focuses on digitizing government services and processes, smart governance goes beyond that to create a more interconnected, data-driven, and citizen-centric governance model. Smart governance uses digitalization as a tool but also incorporates other elements like data analytics, citizen participation, and sustainability to achieve broader governance goals.

Challenges of digital governance

Digital governance, while offering significant opportunities for enhancing public administration and citizen engagement, also presents a range of complex challenges that can impede its successful implementation and sustainability across different contexts. These challenges must be carefully navigated to fully realise the potential of digital transformation in the public sector.

1. Digital divide: One of the most urgent challenges is the digital divide, which signifies the disparity between individuals who have access to digital technologies and those who do not. This divide can manifest in several ways, including disparities in internet access, the availability of digital devices, and the quality of digital infrastructure. In many regions, mainly rural or economically disadvantaged, citizens lack the necessary resources to participate in digital governance initiatives. This exclusion undermines the inclusivity of such programs and exacerbates existing inequalities, limiting the reach and effectiveness of government services.
2. Cybersecurity risks: As governments increasingly depend on digital platforms for service delivery and operational management, they become

more susceptible to cyber threats. Cybersecurity risks, such as hacking, data breaches and ransomware attacks, pose significant dangers to the integrity of government systems and the protection of sensitive information. Ensuring robust cybersecurity measures is essential to maintain public trust, protect national security, and prevent the disruption of critical services. However, the evolving nature of cyber threats requires continuous vigilance, investment, and adaptation of security strategies.

3. Privacy concerns: The digitisation of government services often involves collecting, storing and processing vast amounts of personal data. This raises critical privacy concerns, as citizens may be apprehensive about how their information is used and who has access to it. Governments must balance transparency, which is vital for accountability and privacy protection. Developing clear and strong data protection policies, along with ensuring compliance with privacy regulations such as the General Data Protection Regulation (GDPR) in the European Union, are essential measures for addressing these concerns.
4. Interoperability issues: Effective digital governance requires seamless integration across various digital systems and platforms government departments and agencies use. However, achieving interoperabilit, where these systems can communicate, share data and work together efficiently, poses a significant technical challenge. Disparate legacy systems, varying standards and siloed data repositories can hinder coordination and collaboration, leading to inefficiencies and a fragmented user experience for citizens.
5. Digital literacy: The success of digital governance is heavily dependent on the digital literacy of both the public and government employees. Citizens must possess the skills and knowledge to navigate digital platforms, access services and engage with government initiatives online. Similarly, public sector employees must be proficient in using digital tools and technologies to perform their duties effectively. A lack of digital literacy can lead to underutilising digital services, perpetuating the digital divide and reducing the overall impact of digital governance efforts.
6. Regulatory and legal barriers: The swift progression of technological innovation often outstrips the ability of legal and regulatory frameworks to keep up. Many existing laws and regulations were designed for a pre-digital era and may not adequately address the complexities of digital governance. This can create legal uncertainties and hinder innovation as governments struggle to update or create new regulations that support digital transformation while protecting public interests. Navigating these regulatory challenges requires close collaboration between policymakers, legal experts and technology providers.

7. Change management: Implementing digital governance is not just a technical challenge but also a cultural one. Resistance to change within government organizations can slow down or even derail digital transformation initiatives. Public sector employees may be accustomed to traditional working methods and reluctant to adopt new technologies or processes. Effective change management strategies, including leadership support, training, and clear communication, are essential to overcoming this resistance and fostering a culture of innovation within the public sector.
8. Resource constraints: Digital governance initiatives require significant technology, infrastructure, and human resources investments. For many governments, particularly in developing countries, these resources may be limited, making it challenging to implement and sustain digital transformation efforts. Ensuring that digital governance is both practical and equitable requires careful consideration of resource allocation, strategic planning, and the mobilization of external support, such as international aid or public-private partnerships.
9. Ethical issues: The increasing use of artificial intelligence (AI) and data analytics in governance introduces new ethical dilemmas. These technologies, while powerful, can also raise questions about fairness, accountability, and bias. For instance, AI algorithms involved in decision-making can unintentionally amplify existing societal biases resulting in the unequal treatment of various groups. Addressing these ethical concerns necessitates the creation of clear guidelines and standards for AI use in governance, along with continuous oversight to ensure that these technologies are applied fairly, transparently, and with accountability.

In summary, while digital governance offers immense potential for improving public administration and citizen engagement, it is also fraught with challenges that must be carefully addressed. Governments can better harness the benefits of digital technologies to create more inclusive, secure, and effective public services.

Importance of the recommendations given by the OECD

The OECD's recommendations are vital for helping governments adopt more strategic approaches to technology that foster openness, participation, and innovation. These guidelines are particularly valuable for government agencies aiming to enhance coordination, build stronger capacities and establish frameworks that improve the efficiency of digital technologies in delivering public value and increasing citizen trust. Trust levels in each country are influenced by its unique history and culture.[8] These recommendations enable governments to be proactive, resilient, and future-ready, enhancing public trust through improved service performance and responsiveness, while also garnering public support for innovative governance practices.

Impact

The recommendations to digital governance enable a change in culture within the public sector operations to integrate strategic decisions on digital technologies. Governments have failed to make a complete transition to the new digital environment. It is designed to apply to all member countries and offers valuable guidelines for non-members as well, helping them address challenges and identify best practices related to the development and implementation of digital government strategies. The OECD Council adopted the recommendation on 15 July 2014.

World Bank report on governance

In 2022, the World Bank released a report on citizen-centred Services, introducing a citizen-focused service delivery model aimed at bridging the gap between the public sector and its stakeholders. By minimizing interaction difficulties through process re-engineering, advanced data sharing, simplicity and user-focused design, governments can more effectively and efficiently meet citizens' expectations and needs. Over time, this approach can enhance understanding of citizens' needs and preferences, enabling the development of better policies, programs, and services, particularly for vulnerable groups. A government with a deeper understanding of its citizens can modify its service delivery models to meet the specific needs of different population segments in a secure and reliable manner.

This guide provides valuable insights for practitioners and policymakers involved in modernizing and digitizing services, offering strategies to better integrate citizen-centric approaches. It outlines a four-stage modernisation model, identifying key entry points and offering suggestions for incorporating user feedback throughout service validation, design, re-engineering, digitization, and delivery. Research and practice demonstrate that adopting citizen-centric

8 Public Value encompasses various societal benefits, which may include goods or services satisfying citizens' desires; production choices meeting expectations of justice, fairness, efficiency and effectiveness; properly ordered public institutions; fair and efficient resource distribution; legitimate use of resources for public purposes; and innovation and adaptability to changing preferences and demands.

approaches to e-services can increase citizen satisfaction, improve user experience, and support ongoing adaptation as countries enhance their services.

The successful implementation of citizen-centric reforms requires the commitment of government ministries, departments, agencies (MDAs), policymakers and citizens to actively engage in the process. These reforms involve raising awareness, managing change and engaging stakeholders on both the supply and demand sides. While this may prolong the reform process, the resulting high-quality, error-free e-services can ultimately save significant government time and resources. Recent experiences suggest that design thinking workshops can encourage empathy and shift mindsets towards understanding the needs of service recipients.

Business Process Reengineering (BPR) and digitising services take time and necessitate coordination among MDAs, frontline service providers and users. Legal and regulatory updates may be required to ensure services can be legally delivered online. Though coordinating multi-sectoral services involving several MDAs can be challenging, robust interoperability and data exchange platforms can streamline the process for end users. Migrating to digital services benefits from strong institutional coordination mechanisms. Service reform programs are intricate, involving multiple sectors, institutions, and stakeholders. Effective coordination at the highest government levels is crucial for success, and strong political leadership can remove obstacles and encourage the collaboration needed for multisector service digitization.

Each country will need to customise its approach to implementing citizen-centric service reforms. While this guidebook highlights key steps and considerations, the specific path each country follows will differ. Although there are shared actions that nations can adopt, the implementation and digitization processes will be unique, tailored to the country's specific needs and goals. Actively involving citizens in these reforms can yield substantial benefits for both governments and their people.[9]

Conclusion

Citizen-centric governance is a pivotal instrument for sustainable development, bridging policy frameworks and the lived realities of individuals and communities. Governments and institutions can develop more inclusive, responsive, and

9 World Bank, *Service Upgrade: The GovTech Approach to Citizen-Centered Services* (Washington, DC: World Bank, 2022), <https://thedocs.worldbank.org/en/doc/c7837e4efad1f6d6a1d97d20f2e1fb15-0350062022/original/Service-Upgrade-The-GovTech-Approach-to-Citizen-Centered-Services.pdf>, accessed 25 August 2024.

effective policies that align with the needs and aspirations of the population by placing citizens at the center of governance. This approach ensures that governance is not just about implementing policies but about empowering citizens to actively participate in decision-making that affects their lives and the environment. Integrating citizen participation into governance structures enhances transparency, accountability, and trust in public institutions. It fosters a sense of ownership among citizens, encouraging them to participate actively in sustainable development initiatives. This participatory approach is crucial for addressing complex challenges such as climate change, resource management, and social equity, where the success of policies often depends on the collective action and commitment of the entire community. Citizen-centric governance promotes social cohesion and resilience by ensuring that the voices of marginalised and vulnerable groups are heard and considered. This inclusivity strengthens the social fabric and enhances the legitimacy and sustainability of development outcomes. Citizens take part in the governmental development activities more willingly if they feel that they are a part of the processes.

Inclusive Development

Introduction

Despite the rapid economic growth of the nations globally, poverty reduction has not been relatively proportional. Economic growth does not ensure that it caters to the developmental needs of poor people. It blatantly neglects the issues of inequality and unemployment. The concept of inclusive development emphasises 'pro-poor' equitable economic growth that provides opportunities to all, focusing on equity, physical wellbeing, human competence, ecosystem quality, social welfare and food safety while considering the pace and pattern of development. However, there is a fundamental difference between equality and inclusive development.[1] The former is an outcome, while the latter is more of a process, presumably leading to the former.

Meaning and Definition

Inclusive development does not have a clear definition, but many agencies have attempted to define it by defining its extents and exigent essentials. There have been numerous investigative studies on inclusive development, especially in the and transitional countries Least Developed Countries (LDCs).

Organization for Economic Cooperation and Development (OECD)

Speedy and sustained poverty alleviation is based on inclusive growth in an ecosystem where people are facilitated to add to economic growth on one hand and gain from it on the other.[2] This definition implies that economic growth must have a broad base to ensure that all sectors gain from it to be included in the nation's labour force and market. This definition has the following two dimensions:

- Macro dimension: It denotes the relevance of the essential changes required in the structure to attain economic expansion.

---

1 Dilip Dutta, *Inclusive Growth and Development in the 21st Century: A Structural and Institutional Analysis of China and India* (London: World Scientific Publishing, 2014).

2 United Nations Development Programme, <www.undp.org>, accessed 18 August 2024.

- Micro dimension: The market should be competitive by removing restrictions to growth and widening the investment scope.
- So, there is a direct and undeviating connection between the dimensions of micro and macro determinants of growth.[3]

UNDP

It is crucial to involve diverse groups in creating opportunities, equitable distribution of development gains, and engagement in decision-making processes to effectively address poverty.[4]

World Economic Forum (WEF): The Inclusive Growth and Development Report 2017 outlines the basic features of inclusive growth and development. They are:

- The entire spectrum of the policy framework must be viewed as an integrated system with respect to periodic upgrading of the cultivation system.
- Due importance must be given to balanced fiscal transfers to have broad socioeconomic inclusion.
- Social inclusion must encompass the minutest business and political ethics, tax code, economic integration, intervention of capital investment, high-yielding occupation, eradication of rent lords and quality and equity of education.
- It must be pro-labour and pro-business to boost social inclusion and economic efficiency through a stronger focus on institutions.

There have been reports on country-wise rankings and the SDGs since 2017 from time to time. They have exhaustive data on all the 193 member nations of the UN.

ADB on inclusive development

According to ADB there is no commonly accepted definition of inclusive development. Inclusive development focuses on 'growth combined with equal opportunities'.[5] Its objective is to facilitate equal chances for all and not just the

---

3 OECD, *Growing Unequal? Income Distribution and Poverty in OECD Countries* (Paris: OECD Publishing, 2008).

4 United Nations Development Programme, <www.undp.org>, accessed 18 August 2024.

5 I. Ali and H. Son, 'Defining and measuring growth: Application to the Philippines', *ERD Working Paper Series No. 98* (Manila: Asian Development Bank, 2007).

poor.[6] Inclusive growth occurs when all members of society equally participate in and contribute to the growth process, regardless of their circumstances.

In a keynote speech at the regional conference 'Inclusive Growth and Poverty Reduction in New Asia and Pacific', held on October 8–9, 2007, in Manila, [Walton] highlighted several key points for inclusive development. Walton called attention to the following important points to be considered in inclusive development.[7]

- Focus on inclusion across various economic and social policy areas, taking into account the complementary relationship between equity and economic growth.
- Significant implications for policies aimed at promoting economic empowerment and providing the means to develop human capacities for comprehensive risk management and administration, as well as the administration of justice.
- A wider context in which economic decision-making occurs within existing political processes, group-based interactions, access to information, and accountability mechanisms.

The three pillars of inclusive development according to ADB

According to the Inclusive Economic Development study, three pillars must be interlinked to formulate a comprehensive programme for inclusive development.[8] They are given in Figure 27.

- Inclusive social development pillar: The concept is closely tied to the growth pillar in poverty reduction. This connection highlights how higher education and health indicators contribute to sustainable pro-poor economic growth. Higher education and health indicators contribute to sustainable pro-poor economic growth.
- Economic growth pillar: The economic growth pillar is the second pillar in the poverty reduction strategy and is fundamental for financing the social needs of health and education.

6 I. Ali and J. Zhuang, 'Inclusive growth toward a prosperous Asia: Policy implications', *ERD Working Paper No. 97* (Manila: Asian Development Bank, 2007).

7 M. Walton, 'Poverty reduction in the new Asia and Pacific: Key challenges of inclusive growth for the ADB', *Technical Note* (Manila: Asian Development Bank, 2007).

8 Ganesh Rauniyar and Ravi Kanbur, 'Inclusive growth and inclusive development: A review and synthesis of Asian Development Bank literature', *Occasional Paper No. 8* (Manila: Asian Development Bank, 2009).

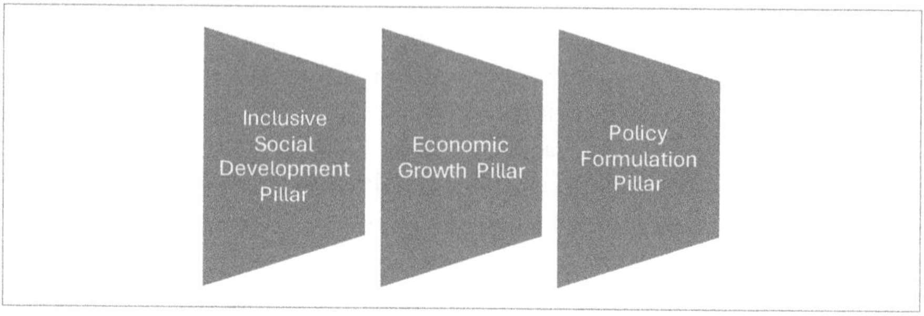

Figure 27 The three pillars

- Policy formulation pillar: The third pillar, policy formulation, involves implementing suitable macro-economic and social protection policies, ensuring accountability in administration, and delivering public services efficiently. Finally, regulatory agencies and good governance can help get closer to the inclusive development goal.

OECD on inclusive development

The OECD defined inclusive development as 'Inclusive Growth is a type of economic growth that ensures opportunities are available to all segments of the population. It aims to distribute the benefits of increased prosperity fairly across society, both in terms of money and other forms of wealth'.[9] It has developed an inclusive growth framework which describes inclusive development instead of defining it.

According to the OECD, the critical features of an 'inclusive governance model' to attain inclusive development are as depicted in Figure 28.

1. Voice: Citizen participation is vital in designing and implementing public policies involving partnerships integrating groups with the help of digitalization. The significant issues to be considered in this feature of voice are as follows:
   - Let everyone have a voice and develop an infrastructure for greater engagement in policy-making and delivering essential services.
   - Better conditions for participation in decision-making for all groups in the country are needed.
   - The conditions must consider gender and broader diversity for outreach.
   - Shun all kinds of lobbying in policy-making through transparency and public consultation.

9 Rolf Adler, *The Governance of Inclusive Growth* (Paris: OECD Publishing, 2016).

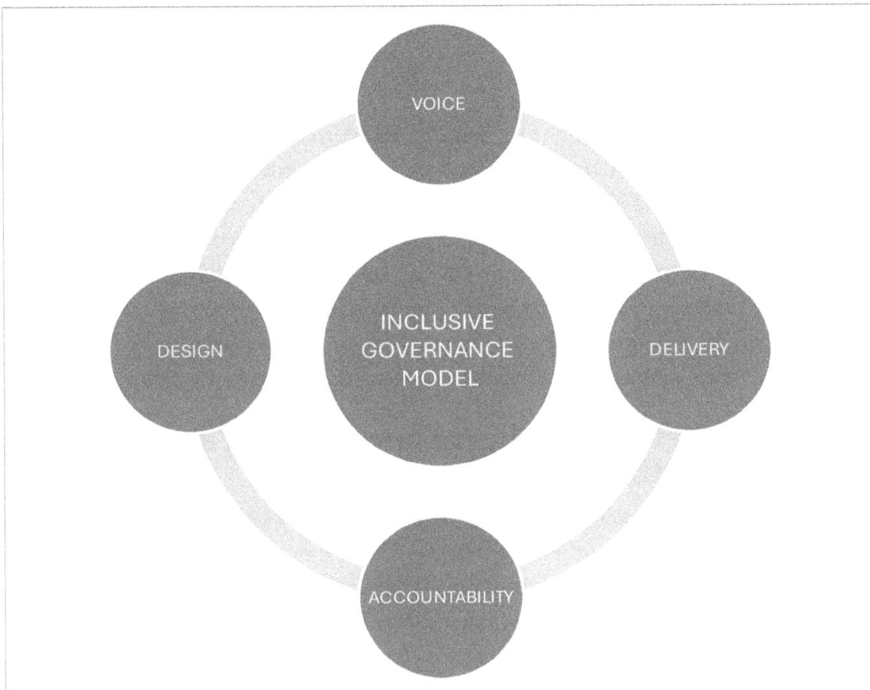

Figure 28 Features of the inclusive governance model

2. Design: The second important feature of inclusive development is designing policies by focusing on their implementation and evaluation. Good governance is critical to consider the multidimensional factors of policy design to include all social groups. The main issues covered in the policy design are the following:
   - All income and non-income dimensions should be included to achieve the desired outcomes.
   - Capacity-building of government agencies must deal with the complex issues involved in public policy-making.
   - Develop tools and processes to align the government's policies.
   - Define the needs and aspirations precisely in the proper manner.

   The government must work cohesively with all its agencies, using all kinds of mechanisms in its power and making effective policies by budgeting based on the determined policies.
3. Delivery: The government must, essentially, work to deliver services related to health, education, justice, and labour, including all the social groups on the list of beneficiaries. The role of the government is paramount in inclusive delivery. The main points covered in this feature are as follows:

- Skill development in new areas, especially ICT.
- Civil services must operate in alignment with inclusive development goals by strong leadership.
- Focus on gender issues and include them in the inclusive fold.
- Inclusive delivery by diverse and inclusive institutions.

4. Accountability: Policy evaluation is vital to ensure accountability in achieving inclusive development. This requires developing a coordinated government-wide evaluation system focusing on performance management. The OECD states, 'Public governance mechanisms and processes can enhance policymakers' capacity to design, implement and evaluate inclusive growth policy packages and the underlying institutional delivery frameworks.'

Differences between inclusive development and inclusive growth

Inclusive development was first mentioned in 1998 but became more widely used from 2008 onwards. Although the terms 'inclusive development' and 'inclusive growth' are often used interchangeably, they differ in their structural characteristics and components. Inclusive growth is defined as economic growth that generates new opportunities and ensures that all segments of society, especially those with low incomes, have equal access to these opportunities. This definition implies focusing on the following results in inclusive growth: increasing the per capita income in economic, and having policymaking directed towards well-being.

Recent data from the United Nations Conference on Trade and Development (UNCTAD), released on 16 November 2023 highlights the limitations of Gross Domestic Product (GDP) as a comprehensive measure of progress. It reveals that increased economic output does not necessarily lead to more inclusive and sustainable growth. The Inclusive Growth Index, introduced in June 2022, evaluates traditional economic metrics like GDP and also considers living conditions, equality, and environmental sustainability factors. In 2023, the index was expanded to include major economies such as China and India, now covering 129 countries that account for 93% of the global population and 96% of the world's GDP. The latest results show persistent disparities, with developed economies achieving nearly twice the average score in inclusive growth compared to developing ones. However, the gap is narrowing in some areas. For example, in environmental performance, developed countries scored an average of 42.5 out of 100, compared to 31.3 for developing countries – a less pronounced difference than in economic performance, where the scores were 41.7 and 14.7, respectively. These findings support UN Secretary-General António Guterres's call to look beyond GDP as the sole measure of progress.[10]

Inclusive development defined

Targeted and intentional growth often results in wealth concentration, excluding many marginalised labour market segments. Inclusive development advocates for direct democracy, which involves exercising civil, civic, and political rights, and the equitable distribution of amenities such as health, education, and infrastructure to ensure everyone can access these resources1. Therefore, the inclusive development approach is an investment in 'renewable capital' to sustain societal well-being. This approach emphasises investing in human well-being, including that of future generations, and managing the ecosystem services that support such well-being, rather than focusing solely on the growth rate.[11]

Inclusive Development Index

In its Inclusive Growth and Development Report 2017 in Davos, Switzerland, the World Economic Forum (WEF) introduced key performance indicators (KPIs) to assess inclusive growth and development. These KPIs illustrate a country's institutional strength in policy formulation and implementation compared to its peers. It has been recognised that GDP alone is insufficient as a measure of inclusive development. Modern economies operate digitally, significantly affecting business operations. Government tools cannot accurately measure transaction costs and prices. The rise in online transactions due to digital consumerism has not been reflected in the GDP. The multidimensional aspect of inclusive development is not captured in its entirety by the measurement of GDP as it cannot include the intangible growth of knowledge-based production instead of capital-based production, nor does it monitor social progress.

The Inclusive Development Index Framework

The Inclusive Development Index framework illustrates the ecosystem of structural policy incentives and institutions that ensure the benefits of expanding national economies are shared by all. This framework outlines a set of indicators for policy and enabling environmental conditions across seven main domains (pillars) and fifteen sub-domains (sub-pillars). These indicators

10 United Nations Conference on Trade and Development, <https://unctad.org/news/unctads-inclusive-growth-index-underscores-need-move-beyond-gdp>, accessed 18 August 2024.

11 I. Sochs, *Inclusive Development Strategy in an Era of Globalization, Working Paper No. 35* (Geneva: Policy Integration Department, World Commission on the Social Dimension of Globalization, International Labour Office, 2004).

describe the range of structural factors within a modern economy that affect the extent of improvements in living standards. The national KPIs display each country's performance on these indicators. The countries are divided into two groups due to different levels of poverty: advanced economies and developing economies.

National Key Performance Indicators (KPIs)

The national KPIs are as follows.

1. Growth and development: The first indicator is growth and development, which comprises the following:
   - GDP per capita
   - Labour productivity
   - Healthy life expectancy
   - Employment
2. Inclusion: Inclusion is comprised of the following:
   - Net-income Gini
   - Poverty rate
   - Wealth Gini
   - Median Income
3. International equity and sustainability and what it supports:
   - Adjusted net savings excluding carbon damage (gross national income)
   - Carbon intensity of GDP
   - Public debt
   - Dependency ratio

Inclusive Development Index-Supporting Indices

The Inclusive Development Index evaluates 109 countries for inclusive development. Among these, 79 countries are classified as developed economies, while 39 are categorised as developing countries. The categorization is aimed at fair and just comparison because their poverty line is different. The inclusion index is supported by other indices, giving a more complete understanding of a society's wellness.

- Human Development Index: The UN evaluates life expectancy, education, and per capita income.
- Social Progress Index: It monitors social and environmental well-being.
- World Happiness Index: It comprises measures of generosity, freedom and corruption.

Conditions necessary for inclusive development

In the present context, development administration has been conceptualised as inclusive development that includes social inclusion besides economic growth. Socioeconomic inclusion is a challenge to policymakers. The following are the conditions necessary for inclusive development.

1. Inclusive innovations: Innovative projects, as highlighted by the OECD in 2015, (OECD, 2015) are designed to directly benefit lower-income and marginalised groups.[12] Inclusive innovations often modify existing technologies, products or services to better meet those groups' needs. For example, the TATA Nano car was introduced to cater to the needs of the lower-income group in India. Another example is China, which produces low-priced goods, especially for low-income people.
2. Education policies: According to the OECD, education policies must be innovative for skill development in all socioeconomic segments. Not-for-profit organizations must be encouraged to participate in innovation, which can be funded through public budgets for financial sustainability.
3. ICT: Advantaged technologies in the financial sectors can help achieve inclusive development, such as flexible regulatory agencies, mobile telephony, credit options, microloans, etc. Promoting science and technology will improve productive capacity and the competitive edge of developing for long-term sustainability.
4. Democratization of innovative inclusiveness: The environment in a society must be free to encourage innovative inclusiveness. Innovation-driven growth will develop skills and capacities covering the entire industrial and territorial fields. Three points are critical in this condition.
   - Skewed value distribution: Innovation ideas are few and become successful, covering a significant market share, resulting in skewed value distribution. Its impact is that the benefits lie in the hands of a few.
   - Small-scale entrepreneurs: ICT has enabled small-scale entrepreneurs to democratise innovation to become successful. It involves widening the group of innovators.
   - Policies: Innovation policies will help create conditions necessary for inclusive development.

12 The bottom 40 per cent (respondents above fifteen years of age in the bottom 40 per cent income bracket having an account in their name or someone else's name in a financial institution).

5. Long-term strategic framework (LTSF): Poverty reduction strategies aimed at inclusive development involve the Long-Term Strategic Framework (LTSF). The Asian Development Bank (ADB)[14] included social development as one of the core strategic areas of intervention, alongside sustainable economic growth. This framework must be supported by a comprehensive social development program that prioritises people and empowers marginalised groups in society to gain access to opportunities. Citizen engagement was another powerful core area for inclusive development stressed by capacity-building and targeting the delivery of essential social services to all.
6. The medium-term strategies I and II: Five-year periods were suggested to achieve LTSF with particular reference to the inclusion of women by investing in economic and social infrastructure. Medium-term strategy I and II covered social protection in the following:
   - Rural development (water management, rural infrastructure and agricultural finance)
   - Crucial social development programs for education, health, and gender equality.
7. LTSF (2008–2020): The ADB proposed LTSF (2008–2020) and laid down certain conditions by which inclusive development can be achieved. They are as follows:
   - Generating and increasing economic opportunities
   - Expanding access to these opportunities

Poor governance must be rectified and market-resistant factors must be removed. The OECD stated, 'Effective governance is necessary for economic prosperity and social cohesion'. Hence, the abovementioned conditions are instrumental in mapping the inclusive development path.

Intersecting paths: The convergence of inclusive and sustainable development

In the quest for global development, two critical paradigms have emerged as essential frameworks: inclusive and sustainable development. Each represents a distinct approach to addressing the multifaceted challenges of the modern world. Inclusive development ensures that economic growth and societal progress are equitable, reaching all parts of society, especially those who are marginalised and vulnerable. It emphasises social justice, aiming to dismantle barriers that perpetuate inequality and exclusion.[13] Conversely, sustainable development

13 Amartya Sen, *Development as Freedom* (Oxford: Oxford University Press, 1999).

is centred on balancing economic growth with environmental stewardship, meeting the needs of the present while safeguarding the ability of future generations to meet their own needs.[14] While these paradigms have traditionally been viewed through separate lenses – one prioritizing social equity and the other environmental sustainability – their goals are deeply interconnected. As the global community increasingly recognises the mutual reinforcement between social inclusion and environmental sustainability, these two paradigms converge, leading to a more integrated and holistic approach to development. This intersection is necessary and a powerful opportunity to create a just and sustainable future for all.

Inclusive development and sustainable development are closely related but distinct concepts within the broader framework of global progress. Inclusive development emphasises equitable growth, aiming to ensure that all segments of society, especially marginalised groups, have access to opportunities and resources. It focuses on reducing inequality, promoting social inclusion, and ensuring that the benefits of economic development are distributed fairly. Key elements include access to education, healthcare, social protection, and economic participation for women, minorities, and disadvantaged communities. The objective is to foster a fairer society by tackling economic, social, and political disparities. On the other hand, sustainable development integrates economic, environmental, and social dimensions, aiming to address current needs while ensuring that future generations can meet their own. While it overlaps with inclusive development in its focus on social equity, sustainable development also prioritises environmental conservation and long-term economic viability. It seeks to balance growth with the safeguarding of natural resources and ecological stability.

The primary similarity between the two is their shared focus on social equity and long-term human well-being. However, the difference lies in their scope – sustainable development is broader, addressing environmental sustainability alongside social and economic issues, while inclusive development specifically targets reducing inequality and promoting social justice. Both are essential for holistic progress.

14 World Commission on Environment and Development, *Our Common Future* (Oxford: Oxford University Press, 1987).

Inclusive development: focus on equity and social justice

In the quest for global progress, two critical paradigms – inclusive development and sustainable development – have emerged as essential frameworks. Inclusive development centres on equity, ensuring that economic growth advantages every segment of society, with a special focus on the marginalised and vulnerable. It addresses social exclusion by dismantling barriers related to poverty, gender inequality, and limited access to essential services, prioritizing enhancing human dignity, agency, and empowerment.[15]

Sustainable development: Balancing economic growth and environmental stewardship

The Brundtland Report describes sustainable development as a way to satisfy current requirements without making concessions for the ability of future generations to meet theirs.[16] This approach highlights the necessity of balancing economic advancement with environmental protection, ensuring long-term sustainability by recognizing the limitations imposed by technology and social structures on the environment's capacity. Additionally, the report underscores the importance of prioritizing the needs of the world's impoverished populations and acknowledging the environmental constraints on growth. While sustainable development has traditionally focused on global environmental and economic policies, its integration of social equity has become increasingly critical, especially given the understanding that widespread poverty, particularly in less developed countries, exacerbates vulnerability to ecological catastrophes.[17]

Sustainability development and poverty

The Brundtland Commission asserted that 'widespread poverty is no longer inevitable'. Sustainable development necessitates addressing the basic needs of everyone and providing opportunities for all to achieve their aspirations for a better life. Poverty remains susceptible to ecological and other disasters. The Commission emphasised two key points: (1) Since the majority of poor populations reside in the least developed countries (LDCs) and developing nations, it is crucial to ensure equity through political systems that involve citizen participation in decision-making; (2) There must be a balance between population size

---

15 Sen, *Development as Freedom*.

16 World Commission on Environment and Development, *Our Common Future*

17 John Elkington, *Cannibals with Forks: The Triple Bottom Line of 21st Century Business* (Oxford: Capstone Publishing, 1997)

and economic growth, aligned with the evolving productive potential of the ecosystem, to achieve sustainable development

Convergence: Mutual reinforcement of social and environmental goals

The United Nations acknowledges that inclusive and sustainable development works in tandem to reduce poverty in all its dimensions.[18] Sustainable development requires economic and inclusive growth, ensuring no one is left behind. This includes the most vulnerable populations within countries and globally, as emphasised by the Sustainable Development Goals (SDGs). Eleven of the seventeen SDGs incorporate elements of inclusive development, aiming to eradicate poverty, eliminate hunger, address gender inequalities, and promote access to potable water, hygiene facilities, power supply, and empowerment.

Integrated approaches and shared challenges

The convergence of these paradigms is evident in the SDGs, which reflect an integrated approach to development. However, challenges remain, particularly in aligning inclusive development with the ecological boundaries necessary for sustainable progress.[19] While inclusive development addresses the micro-level, focusing on specific social issues, sustainable development provides a broader framework that includes ecological considerations. The Sustainable Development Goals (SDGs) emphasise that there should be no trade-offs between economic, social, and ecological objectives. This integrated approach ensures that progress in one area does not come at the expense of another, promoting a holistic and balanced path towards sustainable development. However, achieving this balance requires careful integration of inclusive development principles to ascertain that economic growth delivers true social inclusion without compromising environmental sustainability. Realising the SDGs depends on the success of inclusive development, underscoring the importance of equity and sustainability as intertwined goals in global development efforts.[20]

18 United Nations, *Transforming Our World: The 2030 Agenda for Sustainable Development* (New York: United Nations, 2015).

19 Julian Agyeman, Robert D. Bullard, and Bob Evans, 'Exploring the nexus: Bringing together sustainability, environmental justice and equity', *Space and Polity*, 6.1 (2002), 77–90.

20 W. N. Adger, 'Social and ecological resilience: Are they related?', *Progress in Human Geography*, 24.3 (2000), 347–64.

Table 2 Social equity and inclusiveness

Principles	Sdgs	Achievements and limitations
1. Fair distribution of development opportunities. 2. Knowledge inclusion for all 3. Targeted capacity building for all 4. Protection of the poorest 5. Participation of all in the developmental politics	G1: Resilience to disasters and land ownership G2: Poverty G3: Hunger/Malnutrition G 4: Well-being and healthy lives G4: Inclusive and equitable education G5: Inequality between sexes G6: Water and Sanitation G7: Universal access to energy G8: Employment G9: Inclusive industrialization G 10: Reduce inequalities G 11: Inclusive cities G15: Sustainable livelihoods G16: Inclusive Institutions	• Protection of poor and small-scale sector • Encompasses all development opportunities to a certain degree. • Participation in politics of development is negligible

G implies 'Goals'

Enhancing the principle of inclusiveness in SDGs

The report 'Transforming Our World: The 2030 Agenda for Sustainable Development' enumerates the SDGs, ensuring 'a social floor' and 'planetary boundaries' and the means to achieve them. SDGs cannot be achieved without inclusive development. Green and inclusive growth have additional aspects concerning low-income people's environment or economic share. Inclusive development hinders economic growth's validity as it does not deliver social inclusion. Thus, integrating inclusiveness into the SDGs is essential to ensure sustained and equitable economic growth.

Tables 2, 3 and 4 analyse the dynamics between inclusive and sustainable development.[21]

To conclude, after analysing Tables 2, 3 and 4, it can be said that the SDGs help achieve inclusive development goals to some extent. The most relevant points that sum it up are as follows:

- The SDGs emphasise the element of growth concerning sustainable development.

21 OECD, *Innovation Policies for Inclusive Development: Scaling up Inclusive Innovations* (Paris: OECD Publishing, 2015).

Table 3 Ecological inclusiveness

Principles	Sdgs	Achievements and limitations
1. Develop and adopt ecological limits at all levels 2. Fair Distribution of Rights, Responsibilities, and Risks 3. Enhance resilience and adaptability 4. 'Green' international cooperative mechanisms 5. Engage all stake-holders	G1: Resilience to disasters and land ownership G2: Sustainable agriculture G6: Sustainable water management G7: Sustainable energy G8: Resource efficiency G9: Sustainable industrial-isation G11: Sustainable cities G12: Sustainable consumption and production patterns G13: Combat climate change G14: Sustainably use marine resources G15: Promote sustainable ecosystems and halt land degradation	A few of the goals meet the following: • Ecological limits • Prioritise strengthening resilience and adaptability • Promoting confidence in the financial system. There are no provisions for: • The distribution of rights, responsibilities, and risks among various groups. • The specifics of how international cooperation should be reformed, despite its mention. • Ensuring sufficient stakeholder participation.

Table 4 Relational inclusiveness

Principles	Sdgs	Achievements and limitations
1. Safeguard public and merit goods from privatization or securitization. 2. Tackle all contributing factors and entities; counteract offshoring, tax havens, and other mechanisms that lead to wealth concentration. 3. Challenge narratives that concentrate wealth to ensure markets and growth align with the Inclusive Development framework. 4. Evaluate tools for inclusiveness and ensure accountability to lower levels. 5. Uphold global rule of law and constitutional principles	G3: Tobacco convention with robust relational approach G4: Gender equality G10: Reduce inequality G11: Urban-rural interface G13 &14: Encourages developed countries to take responsibility for climate change and biodiversity G16: Promote peaceful and inclusive societies' access to justice. G17: Revitalise global partnership	'These objectives strive to: • Uphold the rule of law and ensure equal access. • Tackle bribery and corruption. • Eliminate inequality by amplifying voices. However, they fall short in addressing the rationale behind mid-based economic growth, protecting public goods from privatization and securitization, and ensuring accountability at lower levels.'

Figure 29 Inclusiveness and SDGs

- Eleven out of 17 goals concern social inclusiveness for people with low incomes and the most marginalised sectors of society.
- The ecological inclusiveness of SDGs is much lower in relation to their commitment, limited eco-space, and the requirement to improve human welfare.
- Relational politics has the most minor focus on SDGs. However, its efforts to promote private sector responsibility and its emphasis on multi-level decision-making bodies hold the potential for significant changes.
- Most of the goals don't address social, environmental, or relational concerns directly, except for SDG 11, which integrates all three aspects, and SDG 17, which operationalises relational issues, as given in Figure 29 below.

Community empowerment for inclusive development

Few goals address social, environmental, or relational matters comprehensively, with the exception of SDG 11, which incorporates all three aspects, and SDG 17, which translates relational issues into practical action, as given in Figure 7.5.

Developing the rural and marginalised sectors is a major challenge for any government in all social, economic, and ecological aspects. Such a scale of inclusive development requires an integrated effort of all institutions and technologies, but their implementation faces a lot of impediments, such as red-tapeism and corruption. It is suggested by all the major forums of inclusive development that community empowerment can help governmental institutions to address the difficulties associated with inclusive development, which are mainly the following:

- Reduction of poverty
- Employment generation
- Empowerment
- Equity
- Sustainable development of natural resources

- Delivery of basic services

Not only do the villages need inclusive development, but urban areas also need it because there is a wide gap between the slums and the elite high-rise structure population.

Community empowerment

Empowerment and community participation are vital to achieving inclusive development targets. There is a need to involve the citizens to help international, national, and local agencies and governmental and non-governmental agencies address these issues.

Definition

The World Health Organization (WHO) defines 'Community Empowerment' as the process by which groups of individuals gain the ability to take control of their lives.[22] These groups, or communities, may not always be physically connected but share similar interests, identities, or concerns. The primary purpose of community empowerment is to make the communities engage and participate in the process of inclusive development and take ownership and action to renegotiate and gain more control. NGOs have been termed 'social agents' as the researchers work with the local people and agencies. The World Bank recognises community empowerment via SHGs and its development projects, an instrument to ascertain that the benefits are availed by the poorest most efficiently and cost-effectively by sharing costs and benefits. The Brundtland Commission Report emphasised that a key condition for sustainable development is ensuring meaningful and active participation of citizens.

Role of Stakeholders in Inclusive Development

Inclusive development is cultivated by engaging all stakeholders, especially regarding sustainability. They consider the needs of all the beneficiaries while delivering development goods and services. Governments, non-profits, businesses, and community groups each bring something valuable to the process, from resources to local knowledge, all working together to make development more effective and fair.

---

22 Asian Development Bank, *Moving the Poverty Reduction Agenda Forward in Asia and the Pacific: The Long-Term Strategic Framework of the Asian Development Bank (2001–2015)* (Manila: Asian Development Bank, 2001).

Government Role

Governments are primary stakeholders in inclusive development, responsible for creating and implementing policies that promote equity and sustainability. They establish regulatory frameworks that can either facilitate or hinder inclusive development. For instance, policies that promote equitable access to education and healthcare can help bridge gaps between different socioeconomic groups (UNDP, 2020). Governments also play a pivotal role in allocating resources and funding for programs that target underserved populations. Effective governance ensures that these resources are used efficiently and reach the intended beneficiaries.

Civil Society Organizations

Civil society organizations (CSOs) advocate for marginalised groups and hold governments and corporations accountable. They often work on the ground, providing services and support where government programs may fall short. CSOs have a significant critical task in educating people about issues of social equity and inclusiveness, mobilizing public opinion, and influencing policy changes (Oxfam, 2021). Their grassroots involvement helps ensure that development projects reflect the real needs and demands of the areas they want to work for.

Private Sector Involvement

The private sector contributes to inclusive development by integrating social responsibility into their business models. Corporations can drive inclusivity through fair employment practices, equitable supply chains, and investments in local communities. For example, companies that implement diversity and inclusion programs in their workforce can help reduce systemic inequalities (World Economic Forum, 2022). Additionally, businesses can collaborate with other stakeholders to create novel and modern resolutions that address social and environmental challenges, such as sustainable production practices and inclusive financial services.

Local Communities

Local communities are both beneficiaries and active participants in inclusive development. Their involvement is crucial for ensuring that development initiatives are contextually relevant and culturally sensitive. Community engagement helps identify local needs and preferences, which can guide the design and implementation of development programs (World Bank, 2023). Community-based approaches also empower individuals, build local capacity, and foster ownership of development projects, leading to more sustainable and impactful outcomes.

Collaborative Approaches

Effective inclusive development requires collaboration among these diverse stakeholders. Multi-stakeholder partnerships facilitate the sharing of knowledge, resources, and expertise, enhancing the effectiveness of development initiatives. For instance, public-private partnerships (PPPs) can leverage the strengths of both sectors to achieve common goals, such as improving infrastructure or expanding access to services (OECD, 2024). These collaborations also help align interests and address potential conflicts, fostering a more cohesive approach to development.

Thus, stakeholders play interconnected and complementary roles in advancing inclusive development. Governments set policies and provide funding, CSOs advocate and deliver services, the private sector offers resources and innovation, and local communities ensure relevance and participation. A collaborative approach among these stakeholders enhances the potential for achieving sustainable development goals and ensuring that everyone progresses without glaring disparities.

Conclusion

Inclusive development stands at the crossroads of equity and progress, offering a pathway to a more just and equitable society. Inclusive development tackles the underlying reasons of inequality and marginalization by ensuring that all individuals, regardless of their socioeconomic status, gender, ethnicity, or geographical location, have access to economic growth and social progress benefits. It moves beyond the mere pursuit of economic development to embrace the principles of social justice, human dignity, and participatory governance. This approach uplifts the marginalised and strengthens the social fabric, fostering social cohesion and stability.

Integrating inclusive development into global and national policies is essential in today's interconnected world, where disparities can lead to social unrest and hinder collective progress. Policymakers can prevent economic growth from compromising social fairness or environmental sustainability by incorporating inclusive development goals into larger frameworks like the Sustainable Development Goals (SDGs). This alignment helps ensure that progress is both equitable and sustainable. Aligning these frameworks becomes essential for tackling issues like poverty, gender inequality, and climate change, as these challenges often hit vulnerable communities the hardest.

Inclusive development is not just a moral imperative but also an economic one. Societies that prioritise inclusion tend to be more resilient and innovative as they draw on the talents and contributions of all their members. Nations can create a virtuous cycle of growth that benefits everyone, leading to more sustainable and long-term prosperity by investing in education, healthcare, and opportunities for all.

As the global community continues to grapple with complex challenges like poverty, climate change, and social inequities, the principles of inclusive development provide a critical framework for sustainable and resilient societies. We can create a world where no one is left behind, ensuring that the fruits of progress are shared equitably and that future generations inherit a world that is not only prosperous but also just and sustainable by integrating inclusiveness into all aspects of development – economic, social, and environmental. The journey towards inclusive development is ongoing and demands continuous effort, innovation, and commitment from all stakeholders. Only through collective action and a shared vision of equity can we hope to build a future that truly reflects the principles of inclusive development.

Annexure I

Framework of Policy and Institutional Indicators

The WEF gave the following framework of policy and institutional indicators by dividing it into pillars which are given below in brief.

1. The first pillar: education and skills development

This is one of the most elementary indicators to measure the inclusive development of a nation and to compare it with other countries. It has three sub-indicators with further classification.

Access

The WEF labelled the first sub-indicator as access to education which further was measured on the following parameters:

- Mean years of schooling
- Gross pre-primary enrolment
- Net primary enrolment
- Gross secondary enrolment
- Gross tertiary enrolment
- Vocational enrolment
- Availability of high-quality training services
- Gender gap in education

Quality

The quality of education is measured with respect to its extent on a scale of 1 to 7 with '1' being 'not well at all' and '7' being extremely well. The following are the criteria:

- Quality
- Internet access in schools
- Public expenditure in schools (% of GDP)
- Pupils to teacher ratio
- OECD's Programme for International Student Assessment (PISA), which is an average standardised test of performance of 15 years of age
- PISA math score
- Basics in reading comprehension
- Basics in math
- Ease of finding skilled employees
- Quality of vocational training

Equity

Equity in all levels of socioeconomic groups is measured by this indicator. Its sub-indicators are the following:

- Resilient students
- Social inclusion
- PISA reading score
- PISA math score
- Mean years of schooling
- Primary completion rate
- Lower-secondary completion rate
- Upper-secondary completion rate
- Basics in reading comprehension
- Basics in math

2. The second pillar: basic services and infrastructure

The second pillar, according to the WEF, is to assess the basic services and infrastructure in countries in comparison to their peers. The following are the indicators and their sub- indicators with respect to the second pillar.

Basic and digital infrastructure:

The basic and digital infrastructure is evaluated on a scale of 1 to 7 where 1 is a value denoting the 'extremely underdeveloped' among the worst of the underdeveloped among the worst in the world, while 7 representing 'extensive and efficient' among the best in the world. The sub-indicators are as follows:

- Quality of overall infrastructure
- Efficiency of ground transport
- Access to electricity
- Transport infrastructure
- Slum population (urban)
- Dwellings without basic facilities
- Internet users
- Fixed broadband internet subscriptions
- Active mobile broadband subscriptions
- Mobile cellular tariffs, purchasing power parity
- Fixed broadband internet tariffs
- Quality of health care services
- Accessibility of health care services
- Particulate matter concentration (adversely affects health)
- Out-of-pocket health expenses
- Undernourishment
- Inequality-adjusted life expectancy
- Access to improved drinking water
- Access to improved sanitation
- Gender gap in health
- Stringency of environmental regulations
- Indoor air pollution xxiii. Reliability of police services

3. The third pillar: corruption and concentration of rents

The third pillar relates to the corruption and concentration of rents which includes various predetermined indicators in the field of ethics in judiciary, business, politicians and payments as follows:

Business and political ethics

The sub-indicators under this category are given below.

- Judicial independence
- Diversion of public funds

- Irregular payments in tax collection
- Ethical behaviour of firms
- Public trust in politicians
- Irregular payments in public contracts
- Favouritism in the decisions of government officials

Concentration of rents

The WEF identifies the following sub-indicators:

- Regulatory protection of incumbents
- Extent of market dominance
- Intensity of competition
- Land inequity Gini
- Effectiveness of anti-trust policy
- Concentration of banking-sector assets

4. The fourth pillar: financial intervention of real economy investment

The indicators and sub-indicators under this pillar are as follows:

Financial system inclusion

This indicator comprises the following.

- Affordability of financial services
- Gender gap in financial access
- Account at a formal institution
- Account used for business purposes of bottom 40 per cent
- Ease of access to loans
- Financing of small and medium enterprises (SMSE)

Intermediation of business investment:

The sub-indicators are the following:

- Local capital market access
- Venture capital availability
- Domestic credit to private sector by banks
- Bank lending to non-financial corporations (percentage of GDP)
- Private investment in infrastructure (total physical assets and payments as percentage of GDP)
- IPO issuances (small cap)

- IPO issuances (large cap)
- Private R&D expenditure
- Gross fixed capital formation, private sector (percentage of GDP)
- Follow-on issuances (percentage of GDP)
- Corporate bond issuance (percentage of GDP)
- Share turnover ratio (percentage) xiii. Share buyback

5. The fifth pillar: asset-building and entrepreneurship

It concerns itself with asset-building and entrepreneurship in a nation's economy and is compared with other peer countries. The indicators and sub-indicators are as follows:

Small business ownership

This indicator was further subdivided into different sub-indicators.

- New business registered
- Attitudes towards entrepreneurial failure
- Patent Cooperation Treaty filed (percentage of population)
- Time required to start a business
- Cost of starting a business
- Time required to resolve insolvency
- Cost of resolving insolvency
- Cost of enforcing a contract
- Time required to enforce a contract
- Time required to prepare and pay taxes

Home and financial asset ownership

This indicator comprises the following sub-indicators.

- Protection of property rights
- Home ownership rate
- Housing loan penetration
- House price-to-price ratio
- Employee stock ownership
- profit sharing
- private pension assets

6. The sixth pillar: employment and labour compensation: The sixth pillar states that the indicators and sub-indicators required to measure the extent of employment and labour compensation reached by the nation economies.

Economic participation and opportunity

The sub-indicators are the following.

- Female labour force participation
- Unemployment rate
- Youth unemployment rate
- Vulnerable unemployment rate
- Extent of informal economy (undeclared or unregistered activity)
- Country capacity to retain talent
- Social mobility
- Strictness of employment protection
- Unusual hours of work
- Share in temporary employment
- Underemployment rate
- Active labour market expenditure (percentage of GDP)

Wage and non-wage compensation

The sub-indicators are the following.

- Low pay rate
- Gender gap in estimated earned income
- Working poor
- Pay linked to productivity
- Wage dispersion
- Trade union density
- Collective bargaining coverage rate
- Cooperation in labour–employer relations
- Workers' rights
- Cost of childcare
- Maternity leave
- Paternity leave

7. The seventh pillar: fiscal transfers

Lastly, the seventh pillar measures and evaluates the level of inclusive growth and development. They are as follows.

Tax code

This indicator comprises the following sub-indicators:

- Extent and effect of taxation on incentive to work
- Extent and effect of taxation on incentive to invest
- Total tax revenue
- Progressivity index
- Total tax wedge (percentage of labour costs)
- Tax on consumption (goods and services percentage of revenue)
- Tax on property
- Tax on capital (percentage of GDP)
- Tax on inheritance (percentage of GDP)

Social protection

The sub-indicators are as follows.

- Efficiency in public goods and services provision
- Social safety net protection
- Benefit-to-cost ratio (poorest quintile)
- Adequacy of social insurance
- Adequacy of social assistance
- Total social public expenditure
- Coverage of old-age pensions
- Coverage of unemployment insurance
- Progressivity of pensions
- Coverage of health care
- Employment injury coverage (as percentage of the labour force)
- Net pension replacement rate xiii. Unemployment insurance

Gender Equality in Sustainable Development

Introduction

Gender equality is acknowledged as a core human right and an essential element of sustainable development. The relationship between gender equality and sustainability is deeply interconnected, as gender inequalities can exacerbate environmental degradation, economic disparities and social injustices. The inclusion of gender equality in the United Nations Sustainable Development Goals (SDGs) underscores its significance to attain a more just, equitable, and sustainable world. This chapter explores the multifaceted role of gender equality in sustainable development, analyzing how gender-responsive policies and practices can contribute to achieving global sustainability goals. This chapter aims to provide a comprehensive understanding of the importance of integrating gender perspectives into development strategies by examining the barriers, opportunities, and important areas where gender equality intersects with sustainable development.

Theoretical framework

Gender equality and sustainable development are intrinsically linked, with the achievement of one often depending on the other. Sustainable development is broadly defined as development that meets the needs of the present without compromising the ability of future generations to meet their own needs.[1] It encompasses economic growth, social inclusion, and environmental protection, all of which are areas where gender plays a crucial role. Gender equality, on the other hand, refers to the state in which access to rights or opportunities is unaffected by gender. It is a fundamental human right and a precondition for social justice. In sustainable development, gender equality is not only a goal in its own right (as articulated in SDG 5) but also a means of achieving other development outcomes. Studies indicate that societies with greater gender equality often experience higher economic development, improved health outcomes, and more efficient governance. From a theoretical perspective, the intersection

1 World Commission on Environment and Development, *Our Common Future* (Oxford: Oxford University Press, 1987).

of gender equality and sustainable development can be understood through several vital frameworks:

1. Capabilities approach: Pioneered by Amartya Sen and Martha Nussbaum, the capabilities approach emphasises providing individuals with the opportunities and resources to lead lives they value. Gender equality, in this context, is about ensuring that both men and women can participate fully in social, economic, and political life. This approach aligns closely with sustainable development, which seeks to expand people's capabilities without depleting natural resources or harming future generations.[2]
2. Ecofeminism: Ecofeminism highlights the connections between the exploitation of women and the degradation of the environment. It argues that patriarchal structures contribute to both gender inequality and environmental destruction. Ecofeminists advocate for a more holistic approach to sustainable development that addresses environmental issues and gender disparities.[3] This perspective is particularly relevant in areas such as climate change, where women are disproportionately affected but also play a critical role in developing adaptive strategies.
3. Intersectionality: The concept of intersectionality, introduced by Kimberlé Crenshaw, emphasises that gender equality can be only comprehended in relation with other social categories such as race, class, and ethnicity. Sustainable development initiatives must consider these intersecting identities to be inclusive and effective.[4] Intersectionality ensures that the unique experiences of marginalised women, often the most vulnerable to environmental and economic changes, are considered in policymaking.

These theoretical frameworks provide a foundation for understanding the critical role that gender equality plays in sustainable development. Societies can unlock the potential of all individuals to contribute to sustainable economic growth, environmental conservation, and social well-being by promoting gender equality.

---

2 M. C. Nussbaum, *Women and Human Development: The Capabilities Approach* (Cambridge: Cambridge University Press, 2000).

3 M. Mies and V. Shiva, *Ecofeminism* (London: Zed Books, 1993).

4 K. Crenshaw, 'Demarginalizing the intersection of race and sex: A Black feminist critique of antidiscrimination doctrine, feminist theory, and antiracist politics', *University of Chicago Legal Forum*, 1989.1 (1989), 139–67.

Historical context

The historical trajectory of gender equality within the global development agenda reflects a gradual but significant shift from marginalization to centrality. In the early years of international development, gender issues were often overlooked or addressed superficially. Development projects frequently assumed a one-size-fits-all approach, neglecting the different needs and contributions of men and women.

The 1970s were a pivotal decade, characterised by the rise of the Women in Development (WID) approach, which sought to integrate women into existing development processes. However, WID was criticised for treating women as a homogenous group and for failing to challenge the underlying power structures that perpetuated gender inequality.[5] This led to the development of the Gender and Development (GAD) approach in the 1980s, which emphasised the need to address the social relations of gender and to involve both men and women in development initiatives.

The 1990s saw the adoption of the Beijing Platform for Action (1995), a landmark document that recognised gender equality as essential to achieving sustainable development. This was followed by the Millennium Development Goals (MDGs) in 2000, which included a specific goal on gender equality and women's empowerment. However, the limitations of the MDGs in tackling structural inequalities and intersectional issues led to the formulation of the Sustainable Development Goals (SDGs) in 2015, with a more comprehensive and integrated approach to gender equality by the United Nations.[6]

Today, gender equality is widely acknowledged as a cornerstone of sustainable development. The global commitment to gender equality is reflected not only in SDG 5 but also in the mainstreaming of gender across all the other SDGs. This recognition is based on the understanding that sustainable development can be achieved by eliminating the structural obstacles that hinder women and girls from achieving their full potential. Gender equality's contribution to sustainable development can be observed in various real-world examples:

1. Rwanda: Rwanda has made significant strides in gender equality since the 1994 genocide, with women now holding over 60 per cent of parliamentary seats, the highest in the world. This shift has been linked to better governance, increased social cohesion, and progress in sustainable development.

5 C. O. N. Moser, *Gender Planning and Development: Theory, Practice and Training* (London: Routledge, 1993).
6 United Nations, *Transforming Our World: The 2030 Agenda for Sustainable Development* (New York: United Nations, 2015).

Women's leadership in Rwanda has contributed to improved health, education and poverty reduction.[7]

2. Bangladesh: Bangladesh's progress in gender equality, particularly in education and health, has been instrumental in its overall development. The country has achieved near parity in primary education enrolment and has seen significant reductions in maternal and child mortality rates. These achievements have, in turn, supported broader economic growth and poverty reduction.[8]
3. Norway: Norway's commitment to gender equality has been a critical factor in its success as one of the world's most sustainable and prosperous countries. The country's gender-responsive policies, including generous parental leave and child care support, have facilitated high female labour force participation, contributing to economic growth and social welfare.[9]

These examples demonstrate that gender equality is a moral imperative and a practical strategy for sustainable development. Societies can create more resilient, equitable, and prosperous futures by ensuring that women and men have the same chances to engage in and gain from development.

Gender equality and the United Nations 'Sustainable Development Goals (SDGs)

The United Nations Sustainable Development Goals (SDGs), adopted in 2015, represent a global blueprint for achieving a better and more sustainable future for all by 2030. Among these seventeen goals, SDG 5 focuses explicitly on achieving gender equality and empowering all women and girls. SDG 5 is not only a stand-alone goal but also a cross-cutting theme that influences the achievement of many other SDGs. It addresses critical issues such as discrimination, violence, harmful practices, unpaid care work, political participation, and access to reproductive health, all of which are essential for ensuring gender equality.

SDG 5 includes several targets to end discrimination against women and girls. These include the elimination of all forms of violence, such as trafficking and sexual exploitation, and the recognition and value of unpaid care and domestic

7 P. Debusscher and A. Ansoms, 'Gender equality policies in Rwanda: Public relations or real policy?', *Development and Change*, 44.5 (2013), 1111–1134.

8 N. Kabeer, 'Gender equality, economic growth, and women's agency: The "Endless Variety" and "Monotonous Similarity" of patriarchal constraints', *Feminist Economics*, 22.1 (2016), 295–321

9 M. Teigen, 'Gender quotas on corporate boards: On the diffusion of a distinct national policy reform', *Comparative Social Research*, 29 (2012), 115–46

work through public services, infrastructure and social protection policies. It also emphasises the importance of full participation of women in leadership roles across all levels of decision-making in political, economic and public life.

The inclusion of SDG 5 in the global agenda reflects the understanding that gender equality is not just a moral or human rights issue but also a critical component of sustainable development. Gender inequalities hinder progress in all other areas of development. For instance, without gender equality, efforts to reduce poverty (SDG 1), ensure healthy lives (SDG 3) and achieve quality education for all (SDG 4) are likely to fall short. The achievement of gender equality thus has a multiplier effect across other development goals, making it an essential focus for global sustainability efforts.

Interconnections with other SDGs

SDG 5 is deeply interconnected with many of the other SDGs, highlighting how gender equality can drive progress in multiple areas of sustainable development:

1. Poverty Eradication (SDG 1): Gender inequality is both a cause and consequence of poverty. Women and girls are disproportionately affected by poverty due to limited access to resources, education, and employment opportunities. Gender disparities are essential to remove or reduce poverty. For example, empowering women economically through access to financial services, land ownership, and decent work can significantly reduce poverty levels and enhance household welfare.[10]
2. Good Health and Well-being (SDG 3): Gender equality ensures healthy lives and promotes well-being. Women and girls often face health risks related to reproductive health, gender-based violence, and lack of access to healthcare services. SDG 3 can be advanced by resolving these issues and ensuring women can access quality healthcare. For instance, improving maternal health and reducing maternal mortality rates are directly linked to gender equality and the provision of reproductive health services.[11]
3. Quality Education (SDG 4): Education is a powerful tool for achieving gender equality. Guaranteeing that girls receive the same educational opportunities as boys leads to broader social and economic benefits, including improved health outcomes, reduced poverty and increased participation in the labour force. Education supports women and girls to make informed

10 N. Kabeer, *Gender, Labour Markets and Poverty: An Overview* (Geneva: UNRISD, 2015).
11 World Health Organization, *Maternal Mortality* (2019), <https://www.who.int/news-room/fact-sheets/detail/maternal-mortality>, accessed 15 August 2024.

choices about their lives and plays a crucial role in disrupting the cycle of poverty and inequality.[12]

4. Decent Work and Economic Growth (SDG 8): Women's economic empowerment is crucial for sustainable economic growth. When women have equal opportunities to participate in the labour market, economies grow faster and decrease poverty. However, women often face barriers such as wage gaps, limited access to leadership roles, and unpaid care responsibilities. Achieving gender equality in the workplace and promoting policies that support work-life balance is essential for driving inclusive economic growth.[13]
5. Reduced Inequalities (SDG 10): Gender inequality is a form of social inequality that perpetuates income, education, and health disparities. SDG 10 aims to reduce inequalities within and among countries by lowering gender disparities. This initiative will ensure that women have equal access to opportunities and resources and that gender-based discrimination is eliminated.[14]
6. Climate Action (SDG 13): Women are disproportionately affected by climate change due to their dependence on natural resources and their roles in communities, although women are also crucial agents of change in climate action. Empowering women to participate in decision-making processes related to climate change can lead to more effective and sustainable solutions. Gender-responsive climate policies that recognise the unique vulnerabilities and contributions of women are crucial for achieving SDG 13.[15]

These interconnections demonstrate that achieving gender equality is not only essential for the success of SDG 5 but also for the broader SDG agenda. The global community can make significant strides toward a more equitable and sustainable future by integrating gender perspectives into all aspects of sustainable development.

---

12 UNESCO, *Global Education Monitoring Report 2020: Gender Report* (Paris: UNESCO, 2020).

13 International Labour Organization, *World Employment and Social Outlook: Trends for Women 2018 – Global Snapshot* (Geneva: ILO, 2018).

14 OECD, *The Pursuit of Gender Equality: An uphill battle* (Paris: OECD Publishing, 2017).

15 UN Women, *Gender Equality: Women's Rights in Review 25 Years after Beijing* (New York: UN Women, 2020)

LGBTQIA+ as a part of gender empowerment for sustainable development

Sustainable development, which aims to meet the needs of the present without compromising the ability of future generations to meet their own needs, is a multidimensional challenge. It involves social, economic, and environmental aspects, with inclusivity being a core element for its success. Gender empowerment, which traditionally focuses on women's rights and equality, must be expanded to include LGBTQIA+ individuals if sustainable development is to be truly inclusive and equitable. The inclusion of LGBTQIA+ communities in gender empowerment efforts is essential for fostering social justice, reducing inequality, and ensuring that all individuals have the opportunity to contribute to and benefit from sustainable development.

Understanding LGBTQIA+ and gender empowerment

The acronym LGBTQIA+ refers to a broad and diverse range of identities and sexual orientations: Lesbian, Gay, Bisexual, Transgender, Queer or Questioning, Intersex, and Asexual, with the + symbol acknowledging the inclusion of other identities not explicitly represented in the acronym. These communities often face significant discrimination, marginalization, and violence, making it imperative to include them in broader efforts for gender empowerment.

Gender empowerment involves improving access to resources, opportunities, and decision-making power for marginalised groups traditionally excluded due to their gender. Historically, gender empowerment has focused on the rights of women and girls, but sustainable development frameworks must also recognise the importance of empowering LGBTQIA+ individuals. Achieving gender equality means including the full spectrum of gender identities and ensuring that all people, regardless of their sexual orientation or gender identity, can fully participate in social, economic, and political life.

LGBTQIA+ rights and the Sustainable Development Goals (SDGs)

The United Nations' Sustainable Development Goals (SDGs) provide a comprehensive framework for dealing with the world's most pressing challenges, from poverty and inequality to climate change and peacebuilding. Several SDGs explicitly or implicitly relate to the rights and inclusion of LGBTQIA+ individuals, highlighting the importance of their empowerment for achieving sustainable development.

SDG 5, 'Achieve gender equality and empower all women and girls', focuses on eliminating all forms of discrimination and violence against women and girls. The principle of equality embodied in SDG 5 extends beyond the binary understanding of gender. It calls for a more inclusive approach to gender that addresses the unique challenges faced by LGBTQIA+ individuals. Similarly, SDG 10, 'Reduce inequality within and among countries', seeks to address inequalities that are rooted in discriminatory practices, including those based on sexual orientation and gender identity.

SDG 3, 'Ensure healthy lives and promote well-being for all at all ages', is also highly relevant to LGBTQIA+ empowerment. Members of LGBTQIA+ communities often face significant barriers to accessing healthcare, including stigma, discrimination, and violence. This can lead to poor mental and physical health outcomes. Inclusive health policies that cater to the specific needs of LGBTQIA+ individuals are essential for achieving universal health coverage and ensuring that no one is left behind.

Incorporating LGBTQIA+ rights into the SDGs is a matter of social justice and essential for achieving the broader aims of sustainable development. Excluding any group from development processes weakens efforts to create more equitable and resilient societies. Empowering LGBTQIA+ individuals as part of gender equality initiatives strengthens social cohesion, fosters innovation, and drives economic growth by ensuring that all people can contribute to and benefit from development efforts.

The role of LGBTQIA+ empowerment in economic development

LGBTQIA+ inclusion in economic development is crucial for ensuring economies are more resilient, dynamic and inclusive. Discrimination against LGBTQIA+ individuals in the workplace is widespread and results in significant economic losses, both for individuals and societies. Studies have shown that workplace discrimination, unequal access to education, and lack of protection under the law for LGBTQIA+ individuals can lead to lower productivity, reduced innovation and missed economic opportunities. Conversely, fostering a more inclusive environment that allows LGBTQIA+ people to thrive leads to greater economic stability and growth.[16]

Inclusive policies that protect LGBTQIA+ individuals from discrimination and violence are fundamental to building a sustainable economy. In countries where legal protections for LGBTQIA+ people are weak or non-existent, many

16 M. V. L. Badgett, 'The economic cost of stigma and the exclusion of LGBT people: A case study of India', World Bank Group (2014), <https://documents.worldbank.org/en/publication/documents-reports/documentdetail/527261468035379692/the-economic-cost-of-stigma-and-the-exclusion-of-lgbt-people-a-case-study-of-india>, accessed 1 September 2024.

individuals face significant barriers to finding employment, accessing financial services, and fully participating in economic activities. Strengthening legal protections, promoting inclusive hiring practices, and ensuring equal access to education and healthcare are essential steps toward creating economies where everyone can contribute and benefit.

Moreover, LGBTQIA+ empowerment promotes economic diversity and innovation. Diverse workforces that include individuals from a wide range of backgrounds, including LGBTQIA+ individuals, are more likely to foster creativity and innovation. By drawing on diverse perspectives, businesses and organizations can develop new solutions to complex challenges, leading to enhanced productivity and competitiveness. This diversity of thought is critical for achieving sustainable development, as it allows for generating innovative ideas and practices that can drive progress across various sectors.[17]

LGBTQIA+ empowerment and social cohesion

Empowering LGBTQIA+ individuals is also vital for promoting social cohesion, a critical component of sustainable development. Societies that are more inclusive and accepting of diversity are generally more peaceful and stable. Discrimination, exclusion and violence against LGBTQIA+ individuals can lead to social unrest, undermine trust in institutions and weaken social bonds. In contrast, societies that embrace diversity and work to eliminate discrimination are better equipped to address collective challenges and foster a sense of belonging among all citizens.

Social inclusion of LGBTQIA+ individuals also contributes to developing more just and equitable governance systems. When LGBTQIA+ people are empowered to participate in decision-making processes, policies are more likely to reflect the needs and priorities of all citizens, leading to more effective and equitable governance. Inclusive governance structures that respect and protect the rights of LGBTQIA+ individuals contribute to the overall health and resilience of democratic systems, ensuring that all voices are heard and considered in the policymaking process.

Challenges and barriers to LGBTQIA+ empowerment

Despite significant progress in many countries, LGBTQIA+ individuals continue to face numerous challenges that impede their full empowerment and inclusion in sustainable development. In many parts of the world, discriminatory laws,

17 OECD, *LGBT Inclusion in the OECD: A Call for Better Data* (Paris: OECD Publishing, 2020), <https://www.oecd.org/social/lgbt-inclusion-2020.pdf>,accessed 1 September 2024.

social norms and cultural practices persist, limiting the ability of LGBTQIA+ individuals to live freely and safely. In some countries, homosexuality remains criminalised, and transgender individuals face significant legal and social barriers to accessing healthcare, education, and employment. Violence and discrimination against LGBTQIA+ individuals are also pervasive in many contexts, with hate crimes, harassment and exclusion from social services being everyday experiences. These forms of discrimination undermine the ability of LGBTQIA+ individuals to contribute to and benefit from sustainable development initiatives, perpetuating cycles of poverty, inequality and marginalization.

To address these challenges, governments, civil society and international organizations must work together to promote legal reforms, raise awareness and create safe and inclusive environments for LGBTQIA+ individuals. Strengthening legal protections, increasing representation of LGBTQIA+ people in leadership positions, and promoting inclusive social norms are key steps toward overcoming these barriers and ensuring that LGBTQIA+ individuals can fully participate in sustainable development. Thus, LGBTQIA+ empowerment is an essential component of gender empowerment for sustainable development. Society can foster greater inclusivity, innovation and resilience by ensuring that LGBTQIA+ individuals are included in social, economic and political processes. Empowering LGBTQIA+ individuals is a matter of social justice and a key driver of economic development, social cohesion and effective governance. For sustainable development to be truly inclusive and equitable, LGBTQIA+ rights must be integrated into all development planning and implementation aspects. Addressing the challenges and barriers faced by LGBTQIA+ individuals is critical for building more just, inclusive, and sustainable societies where everyone has the opportunity to thrive.

Global progress and challenges

While significant progress has been made in advancing gender equality since adopting the SDGs, many challenges remain. Globally, gender gaps persist in areas such as economic participation, political representation, and access to education and healthcare. For example, the World Economic Forum's Global Gender Gap Report 2021 estimates that it will take over 135 years to close the global gender gap at the current rate of progress.[18] The challenges in achieving gender equality are:

---

18 World Economic Forum, *Global Gender Gap Report 2021* (Geneva: WEF, 2021).

- Social norms: One of the significant challenges in achieving gender equality is the persistence of deep-rooted social norms and cultural practices that discriminate against women and girls. These norms often manifest in gender-based violence, early and forced marriages and the undervaluing of women's work, both in the household and the labour market. These issues require policy changes and a transformation in societal attitudes and behaviours.
- Unequal distribution of unpaid care work: It disproportionately falls on women. This limits their participation in the labour market and public life. The COVID-19 pandemic has exacerbated these challenges, as women have borne the brunt of increased care responsibilities, job losses, and health risks.[19]
- Data and monitoring: In addition to these challenges, there is a need for better data and monitoring systems to track progress on gender equality. Many countries lack comprehensive data on gender disparities, making it difficult to assess progress and identify areas that require targeted interventions. Strengthening gender-disaggregated data collection and analysis is essential for ensuring accountability and measuring the impact of policies to achieve gender equality.

Despite these challenges, there are also opportunities for accelerating progress. The global momentum for gender equality has been strengthened by international movements such as #MeToo and campaigns for equal pay and reproductive rights. These movements have brought greater visibility to gender issues and have pushed governments, businesses and civil society to take more concrete actions toward gender equality. The integration of gender equality into the broader sustainable development agenda provides a framework for understanding and dealing with these challenges in a holistic and coordinated manner. The global community can create collaborations that drive progress on multiple fronts simultaneously by leveraging the interconnections between SDG 5 and other SDGs.

Gender equality and climate change

Climate change affects everyone, but its impacts are not felt equally. Women, especially in developing countries, are disproportionately affected due to their social roles, responsibilities and access to resources. Gender inequalities, such as limited access to education, healthcare and economic opportunities, exacerbate women's vulnerability to climate change. Women are often responsible for securing water, food and energy for their families, tasks that become increasingly

19 UN Women, *The Impact of COVID-19 on Women* (New York: UN Women, 2021).

difficult in the face of climate change. For instance, prolonged droughts and unpredictable weather patterns force women to travel longer distances to find water and fuel, increasing their exposure to health risks and gender-based violence.[22] Additionally, as primary caregivers, women are often on the front lines of dealing with the health impacts of climate change, such as malnutrition and the spread of diseases.

In many rural communities, women rely on natural resources for their livelihoods, such as agriculture, fishing and forestry. Climate change disrupts these livelihoods, leading to loss of income and increased poverty among women. For example, in sub-Saharan Africa, where women constitute a large proportion of the agricultural workforce, changing rainfall patterns and soil degradation threaten food security and economic stability.[23] Moreover, women's limited access to resources such as land, credit and technology further hinders their ability to adapt to climate change. In many parts of the world, women are less likely than men to own land, which restricts their ability to implement climate-resilient agricultural practices or to receive compensation in the event of land degradation or natural disasters. Similarly, women's limited access to financial resources and technology constrains their ability to invest in adaptive measures, such as improved seeds or irrigation systems.

Despite these challenges, women also play a crucial role in climate change adaptation and mitigation. Their traditional knowledge and expertise in managing natural resources make them valuable contributors to community resilience. For example, in many Indigenous communities, women have developed sustainable practices for managing forests, water, and soil, which can be vital for climate change adaptation.

Gender-responsive climate policies

There is a growing recognition of the need for gender-responsive climate policies. These policies integrate gender considerations into climate action, ensuring everyone's needs and contributions are considered in climate strategies irrespective of gender or orientation.

Gender-responsive climate policies involve several key components:

1. Inclusive participation: Women must have a voice in climate-related decision-making processes. Involving women in the development and implementation of national climate strategies, as well as in local adaptation planning, will lead to positive outcomes. Women's participation can lead to more effective and equitable climate solutions as they bring unique perspectives and knowledge to the table. Engaging with women at the

grassroots level will inform the policymakers of the ground realities and issues women face daily.

2. Capacity building: Strengthening women's capacities to adapt to climate change is essential. Women must be provided with access to education, training and resources that enable them to implement sustainable practices and adapt to changing environmental conditions. For instance, capacity-building programs that teach women about sustainable agriculture, renewable energy and disaster risk reduction can empower them to contribute to climate resilience.[24]
3. Gender-sensitive data and analysis: Collecting and analyzing gender-disaggregated data is crucial for understanding the differential impacts of climate change on men and women. This data can better inform policymakers on designing climate policies that address women's needs and vulnerabilities. For example, gender-sensitive vulnerability assessments can help identify the groups most at risk from climate change and target interventions accordingly.[25]
4. Access to resources: Ensuring that women have equal access to resources such as land, finance and technology is vital for their ability to adapt to climate change. Gender-responsive policies should eliminate barriers to women's access to these resources and promote equitable distribution. For instance, programs that provide microfinance or land rights to women can enhance their capacity to invest in climate-resilient practices.[26]
5. Protection from climate-related risks: Gender-responsive policies should also consider the specific risks that women face due to climate change, such as increased vulnerability to gender-based violence during disasters. Gender considerations must be integrated into disaster risk reduction and response strategies, prioritising women's safety and well-being. Gender-responsive climate policies not only protect women's rights but also enhance the overall effectiveness of climate action. These policies contribute to more sustainable and inclusive climate solutions by recognizing and leveraging women's roles as agents of change.

Several countries have made progress in implementing gender-responsive climate policies:

1. Bangladesh: Bangladesh has integrated gender considerations into its national climate policies, recognizing the disproportionate impact of climate change on women. The country's Climate Change Strategy and Action Plan includes specific measures to address women's vulnerability and promote their participation in climate adaptation. For example, the plan supports

women's access to climate-resilient livelihoods, such as aquaculture and handicrafts, and ensures that women are involved in community-based adaptation planning.

2. Kenya: Kenya has adopted a gender-responsive approach to climate change through its National Climate Change Action Plan. The plan emphasises the importance of gender equality in climate action and includes provisions for women's participation in decision-making and access to resources. For instance, the plan supports women's involvement in renewable energy projects, such as solar energy initiatives, which provide clean and affordable energy to rural communities while empowering women economically.
3. Philippines: The Philippines has implemented gender-responsive disaster risk reduction policies that recognise the specific needs of women in disaster-prone areas. The country's policies ensure that women are involved in disaster preparedness and response planning and that their safety and well-being are prioritised during emergencies. For example, the government has established women-friendly spaces in evacuation centres to provide support and services to women and girls affected by disasters.

These case studies illustrate the importance of integrating gender considerations into climate policies and the positive impact that such policies can have on both gender equality and climate resilience.

Global and regional initiatives

In addition to national efforts, various global and regional initiatives promote gender equality in climate action. For example:

1. The UNFCCC Gender Action Plan: The United Nations Framework Convention on Climate Change has adopted a Gender Action Plan to enrich the role of women in climate action. The plan aims to improve the gender balance in UNFCCC processes and to integrate gender considerations into national climate policies and strategies. It also supports capacity-building and knowledge-sharing on gender and climate change.[27]
2. The African Union Gender Strategy: The African Union has developed a Gender Strategy focusing on gender-responsive climate action. The strategy promotes women's leadership in climate governance, supports gender-sensitive climate policies and advocates for including women's voices in climate negotiations at the regional and international levels.
3. The Green Climate Fund: The Green Climate Fund, established to support developing countries in combatting climate change, has adopted a Gender

Policy to ensure that its funding mechanisms are gender-responsive. The policy requires all funded projects and programs to include gender considerations, promote women's empowerment and ensure equitable benefits for men and women.

These initiatives demonstrate the global commitment to integrating gender equality into climate action and highlight the importance of collaborative efforts to address the gendered impacts of climate change.

Gender equality and economic development

Gender equality is essential for promoting sustainable economic development, as equal access to opportunities, resources and decision-making for both women and men strengthens economies. Research indicates that reducing gender disparities in education, employment, and political participation leads to greater economic productivity and improved social outcomes. Gender equality supports poverty reduction efforts, fosters equitable wealth distribution, and improves governance. Significant economic potential is unlocked by reducing or removing systemic barriers. Empowering women is conducive to balanced and sustainable growth that aligns with global development goals. Below are the areas that need advancement so that women are on an equal footing.

1. Access to education and skills development: Education and skills development are integral to sustainable development because they directly impact economic growth and social inclusion. Educated women are more likely to participate in the workforce, start businesses, and contribute to economic growth. Moreover, education promotes social equity, a core principle of sustainable development. Communities can improve productivity and foster innovation, leading to sustainable economic progress by enhancing women's educational attainment and vocational skills.[28]
2. Financial inclusion: Financial inclusion is a strategic driver of sustainable economic development. When women have access to financial resources such as credit and savings, they can invest in businesses, improve their household's economic conditions, and lessen poverty. Financial empowerment supports sustainable development by promoting economic resilience and reducing inequality. For example, microfinance and digital financial services can help women build sustainable livelihoods and support community development initiatives, aligning with sustainable economic growth and social inclusion goals.[29]

3. Employment and labor market participation: Increasing women's participation in the labour market supports sustainable development by maximizing human capital and promoting economic growth. Gender equality in employment ensures that all individuals can contribute to and benefit from economic opportunities. Policies that promote equal pay, support work-life balance and eliminate discrimination not only advance gender equality but also enhance overall economic productivity and stability. Nations can achieve more equitable and sustainable economic development by creating inclusive workplaces and reducing gender disparities.[30]
4. Women in leadership and decision-making: Women's representation in leadership and decision-making roles is crucial for sustainable development because it safeguards diverse perspectives in policymaking and governance. Inclusive leadership can lead to more equitable and effective policies that address social, economic, and environmental challenges. Societies can improve governance, foster innovation, and ensure that development strategies are inclusive and responsive to the needs of all citizens, thus supporting sustainable development goals by promoting gender equality in leadership positions.[31]
5. Gender-based violence: Combating gender-based violence (GBV) is essential for sustainable development because GBV undermines women's health, productivity, and economic participation. Reducing GBV supports social inclusion and human rights, which are fundamental to sustainable development. Implementing legal reforms, support services, and awareness campaigns allows communities to create safer environments for women, enabling them to participate fully in economic and social life. Reducing gender-based violence (GBV) contributes to more equitable and sustainable communities.

Case studies and global initiatives

1. The Nordic model: The Nordic model's focus on gender equality aligns with sustainable development by promoting social equity, economic growth, and environmental sustainability. Policies such as parental leave and affordable childcare support women's economic participation. They augment family well-being and work-life balance, contributing to broader social sustainability.[31]
2. Chile's Women's Empowerment Program (WEP): Chile's Women's Empowerment Program supports sustainable development by improving women's access to financial resources, education and employment. This holistic approach enhances women's economic opportunities and

contributes to poverty reduction, aligning with sustainable development goals related to economic growth and social inclusion.[20]

3. India's Self-Employed Women's Association (SEWA): SEWA's model of empowering women through microfinance and vocational training contributes to sustainable development by fostering economic resilience and community development. SEWA's approach supports sustainable livelihoods and reduces poverty, aligning with economic and social sustainability goals.[21]

The OECD's Equal Pay Day highlights the gender wage gap and advocates equal pay for equal work. It raises awareness that women, on average, still earn less than men for similar roles and work. The day promotes policy measures, workplace reforms, and societal change to close this wage gap, emphasizing the economic benefits of gender equality in the workforce. The OECD also encourages governments and businesses to adopt fair pay practices, ensuring that women and men are compensated equally for their contributions. Several countries and other organizations observe their versions of Equal Pay Days to raise awareness of the gender wage gap:

- International Equal Pay Day (September 18) – Established by the United Nations in 2020, this global event emphasises the need for equal pay for work of equal value and gender equality in the workforce.
- Equal Pay Day USA – Held in the US each year, this day symbolises how far into the year women must work to earn what men earned in the previous year. It is typically observed in March or April, with dates varying based on race and ethnicity, reflecting wage disparities among different groups of women.
- European Equal Pay Day – Recognised by the European Union in November, it marks the day when women symbolically stop earning compared to men due to the gender wage gap.
- Equal Pay Day UK – Celebrated in November, it highlights when women effectively start 'working for free' compared to their male counterparts for the rest of the year.

Each event aims to drive attention toward closing the gender wage gap. Hence, societies can advance sustainable development by promoting inclusive growth,

---

20 Chile Ministry of Social Development, *Programa de Empoderamiento de la Mujer* (Santiago: Chile Ministry of Social Development, 2019)

21 SEWA, *Self-Employed Women's Association Annual Report* (Ahmedabad: SEWA, 2020)

reducing inequality, and fostering social cohesion by integrating gender equality into economic development strategies. These case studies and global initiatives highlight the importance of alleviating gender disparities in economic empowerment. Globally, women makeup 39 per cent of the labour force but only hold 28 per cent of managerial positions. In terms of wage equality, women earn 20 per cent less than men on average, with the gender pay gap persisting across sectors. Women also face restricted access to financial services, owning less than 10 per cent of land despite comprising about 43 per cent of the agricultural workforce. Ensuring equal opportunities and financial inclusion for women is vital to boosting economic growth and reducing poverty.[22] Societies can advance gender equality and contribute to more sustainable and inclusive economic development by promoting equal pay, supporting women's participation in the workforce, and expanding access to financial services.

Gender equality and health

Gender equality and health are closely linked within the Sustainable Development Goals (SDGs) framework. SDG 5, which aims to achieve gender equality and empower all women and girls, addresses issues such as eliminating discrimination, violence and disparities in access to education, healthcare, and economic opportunities. SDG 3, focused on ensuring healthy lives and promoting well-being for all, emphasises universal healthcare access, reducing maternal mortality, combating diseases and improving mental health, particularly for women and girls. These two goals are deeply interconnected, as gender equality is essential for improving health outcomes and achieving sustainable development for all. Gender inequality significantly impacts health outcomes and access to healthcare services. Women and men experience health issues differently due to biological differences, but social and economic factors also play a crucial role. Gender inequalities must be brought down in health to achieve sustainable development, as it ensures that all individuals have equitable access to healthcare and can lead healthy lives. Some principal areas in the health sector are:

1. Reproductive health: Reproductive health is a fundamental area where gender disparities are evident. Women's access to reproductive health services, including contraception, prenatal care and safe childbirth, is elemental for their well-being and empowerment. Lack of access to these services can lead to high maternal mortality rates, unplanned pregnancies

22 World Bank, *Women, Business, and the Law 2022*, <https://wbl.worldbank.org>,accessed 31 August 2024.

and adverse health outcomes. Investments in reproductive health services improve women's health and contribute to broader development goals. For example, family planning programs help reduce maternal and infant mortality rates, enable women to pursue education and employment and contribute to economic stability. Ensuring access to reproductive health services aligns with sustainable development goals related to health, gender equality, and poverty reduction.

2. Gender-based health risks: Gender-based health risks, such as higher susceptibility to certain diseases and health conditions, are influenced by biological and social factors. For instance, women are more likely to experience certain types of cancers, such as breast and cervical cancer, while men may face higher risks of cardiovascular diseases. Tackling these health risks associated with gender necessitates customised healthcare solutions and targeted awareness initiatives. For example, screening programs for breast and cervical cancer, along with public health campaigns, can improve early detection and treatment, thereby reducing mortality rates. Such efforts contribute to achieving the sustainable development goal of ensuring healthy lives and promoting well-being for all.
3. Healthcare services availability: Access to healthcare services is a fundamental determinant of health outcomes. Women, particularly in low-income and rural areas, often face barriers to accessing quality healthcare, including financial constraints, lack of transportation, and limited availability of services. Overcoming these obstacles is essential for attaining health equity and promoting sustainable development. Policies that improve healthcare infrastructure provide financial assistance for healthcare costs and ensure the availability of services in underserved areas can enhance women's access to essential health services. For example, mobile health clinics and telemedicine services can improve access for women in remote areas, contributing to better health outcomes and supporting the broader goal of universal health coverage.
4. Mental health and gender: Mental health is an increasingly recognised aspect of overall health, yet it often receives less attention than physical health. Gender differences in mental health, such as higher rates of depression and anxiety among women, are influenced by various social and economic factors, including gender roles, discrimination, and violence. Gender-specific mental health issues are essential to consider in promoting overall mental well-being, which is vital for achieving sustainable development. Mental health services that are sensitive to gender differences must be made accessible while engaging with the stigma and creating supportive environments. For example, integrating mental health services into

primary healthcare and training healthcare providers on gender-sensitive approaches can improve mental health outcomes and support overall well-being.

5. Gender and Non-Communicable Diseases (NCDs): Non-Communicable Diseases (NCDs), such as diabetes, cardiovascular diseases and respiratory conditions, are a growing global health challenge. Gender plays a role in the prevalence, risk factors and outcomes of NCDs. For instance, women may face different risk factors for cardiovascular diseases compared to men, and gender norms can influence health behaviours and access to care. Combating gender disparities in non-communicable diseases requires the implementation of specific prevention and treatment strategies, the promotion of healthy lifestyles and the guarantee of fair access to healthcare services. Public health initiatives that focus on gender-specific risk factors and provide education on NCD prevention can contribute to reducing the burden of these diseases and improving health outcomes.

Case Studies

1. Maternal health programs in Kenya: Kenya has implemented several maternal health programs to reduce maternal mortality and improve reproductive health services. Initiatives like the Beyond Zero Campaign focus on providing healthcare access, training healthcare workers, and improving maternal and neonatal care. These programs contribute to achieving the sustainable development goals related to health and gender equality by reducing maternal and infant mortality rates and promoting women's health.
2. Women's health initiatives in Sweden: Sweden's comprehensive healthcare system focuses on gender-sensitive health services, including reproductive health and mental health support. The country's policies ensure access to free or subsidised healthcare services, including preventive care and mental health support. Sweden's approach to integrating gender considerations into healthcare services contributes to overall health equity and aligns with sustainable development goals.[23]
3. India's National Health Mission: India's National Health Mission includes initiatives to improve maternal and child health, expand access to healthcare services, and address gender disparities in health. Programs like the Janani Suraksha Yojana (JSY) provide financial incentives for institutional

23 Swedish Institute, *Sweden's Health Care System* (Stockholm: Swedish Institute, 2020), <https://si.se/en/education/swedens-health-care-system>,accessed 31 August 2024.

deliveries to reduce maternal and infant mortality rates. These efforts support sustainable development by promoting health equity and improving access to essential health services.[24]

These case studies demonstrate the connection between gender equality in health and sustainable development. Societies can achieve better health outcomes and advance the broader goals of sustainable development by improving access to healthcare services and promoting gender-sensitive health interventions, which reduce gender disparities in health.

Gender equality and education

Gender equality in education is essential for fostering inclusive and sustainable development. When both girls and boys have equal access to quality education, it empowers individuals and strengthens communities and economies. Educating girls has a multiplier effect, leading to better health, higher incomes, and increased participation in decision-making processes. However, gender disparities in education persist in many parts of the world, with barriers such as poverty, cultural norms and early marriage disproportionately affecting girls. Promoting gender equality in education is central to reducing inequality, breaking cycles of poverty, and achieving global development goals. The most critical aspects of education in gender equality are given below:

1. Educational access: Access to education is a fundamental human right and a basic component of sustainable development. Gender disparities in education persist in many parts of the world, with girls often facing barriers such as cultural norms, economic constraints, and lack of infrastructure. These barriers can limit girls' learning and personal development opportunities, impacting their future economic prospects and social participation. Ensuring equal access to education for all genders is essential for achieving sustainable development goals. Policies that promote gender parity in education, such as scholarships for girls, investments in school infrastructure, and community awareness programs, can help overcome these barriers and contribute to broader social and economic development.
2. Quality of education: The quality of education is a critical factor in achieving gender equality. Girls and boys may experience differences in the quality of education they receive, influenced by factors such as teacher training,

24 Ministry of Health and Family Welfare, *National Health Mission* (New Delhi: Government of India, 2021), <https://nhm.gov.in>,accessed 31 August 2024.

curriculum content and school resources. Inadequate education quality can affect students' learning outcomes and future opportunities. Improving the quality of education involves dealing with these disparities by ensuring that all students have access to well-trained teachers, relevant and inclusive curricula and adequate learning materials. Enhancing the quality of education for girls can lead to better educational outcomes, more significant economic opportunities, and more equitable social development.

3. Gender-sensitive curriculum and teaching practices: Gender-sensitive curricula and teaching practices are essential for promoting gender equality in education. Traditional curricula and teaching methods may reinforce gender stereotypes and limit students' opportunities to explore interests and careers outside conventional gender roles. Incorporating gender-sensitive approaches into education can help challenge these stereotypes and promote more equitable learning environments. Gender-sensitive education involves designing curricula that reflect diverse perspectives, acting on gender biases in teaching materials and training teachers to use inclusive teaching practices. These measures can support both girls and boys in reaching their full potential and contribute to achieving gender equality in education.
4. STEM education and gender: Science, Technology, Engineering and Mathematics (STEM) education is crucial for preparing students for the modern workforce and decreasing skills gaps in these critical fields. However, gender disparities persist in STEM education, with girls often underrepresented in these subjects and careers. Promoting gender equality in STEM education involves creating supportive learning environments, providing role models and mentorship for girls, and addressing barriers discouraging girls from pursuing STEM subjects. Initiatives that encourage girls' participation in STEM can contribute to closing gender gaps in these fields and supporting broader economic and technological advancements.[25]
5. Education and economic empowerment: Education is closely linked to economic empowerment, providing individuals with the skills and knowledge needed to participate in the labour market and pursue economic opportunities. Education can improve job prospects, earnings and economic independence for women and girls. Investing in education for women and girls is essential for achieving economic development and gender equality. Policies that support access to education, such as vocational training programs, financial literacy education, and career counselling, can enhance

25 OECD, *The State of STEM Education* (Paris: OECD, 2020)

women's economic opportunities and contribute to sustainable economic growth.

Case Studies

- Girls' education in Rwanda: Rwanda has made significant progress in promoting girls' education through policies like the Gender Monitoring Office and programs like the Girl Effect initiative. These efforts have increased girls' enrolment and completion rates in primary and secondary education, contributing to gender equality and sustainable development.
- STEM education initiatives in India: In India, initiatives such as the STEM for Girls program aim to increase girls' participation in STEM fields by providing mentorship, resources, and hands-on learning opportunities. These programs have shown promising results in enhancing girls' interest and performance in STEM subjects.
- Education for all in Bangladesh: Bangladesh's Education for All program focuses on improving access to quality education for all children, particularly emphasizing girls and marginalised communities. The program has contributed to higher enrolment rates and better educational outcomes for girls, supporting broader gender equality goals and sustainable development.

These case studies demonstrate the importance of lowering gender disparities percentages in education to achieve sustainable development goals. Women's empowerment, particularly in economic and political spheres, has seen progress but remains uneven globally. The promotion of gender-sensitive curricula and supporting girls in STEM can advance gender equality. These efforts contribute to more equitable and sustainable development outcomes by improving access to quality education.

Gender equality and political participation

Gender equality in political participation is essential for creating inclusive governance and fostering sustainable development. When women and marginalised gender groups are equally represented in political decision-making, policies become more inclusive and responsive to diverse societal needs. However, women remain significantly underrepresented globally, holding only 26 per cent of parliamentary seats as of 2023.[26] Gender quotas in more than 130 countries have improved representation. Still, barriers such as discriminatory social norms,

---

26 Inter-Parliamentary Union, 'Women in national parliaments' (2023), <https://data.ipu.org/women-in-politics>,accessed 5 September 2024.

limited access to political networks, and violence against female candidates persist. Increasing gender equality in political participation ensures diverse perspectives shape policies, leading to more equitable and democratic governance.

1. Women in political leadership: Women's representation in political leadership positions is crucial for advancing gender equality and promoting more inclusive governance. Despite progress in some countries, women remain underrepresented in political offices worldwide. This disparity can result in policies that do not fully address the needs and concerns of women and marginalised communities.[27] Increasing women's political representation involves implementing policies such as gender quotas, supporting women candidates and fostering inclusive political environments. Societies can ensure more equitable policymaking and contribute to sustainable development goals by promoting women's political leadership.
2. Gender quotas and legislative changes: Gender quotas are a common strategy to increase women's representation in political positions. Quotas can be implemented at various levels, including national parliaments, local governments, and political party lists. Evidence suggests that gender quotas can increase female representation and influence policy outcomes. Different types of quotas include legislative quotas, which mandate a minimum percentage of women in elected bodies, and voluntary party quotas, where political parties commit to a certain proportion of female candidates. These measures can help address gender imbalances and support more inclusive political systems.
3. Barriers to political participation: Women face several barriers to political participation, including socio-cultural norms, lack of access to resources, and discriminatory practices. These barriers can limit women's ability to engage in political processes and hold leadership positions. These barriers can be addressed by promoting gender-sensitive policies, providing training and support for women candidates and creating an enabling environment for women's political engagement. Overcoming these obstacles can enhance women's political participation and contribute to more equitable governance.[28]
4. Gender and policy impact: Women's participation in political decision-making can lead to more inclusive and responsive policies. Research shows

27 *International Labour Organization, Women and Work: A Review of Progress* (Geneva: ILO, 2019).

28 International Labour Organization, *Global Wage Report 2021: The Gender Pay Gap* (Geneva: ILO, 2021).

that female legislators are more likely to prioritise issues such as health, education and gender equality, which can positively affect social and economic outcomes. Societies can ensure that diverse perspectives are considered in policymaking by increasing women's representation in political institutions. This can lead to more effective and equitable solutions to social challenges and support sustainable development goals.

Case studies and best practices

1. Gender quotas in Rwanda: Rwanda has implemented gender quotas that require women to hold at least 30 per cent of seats in parliament. As a result, Rwanda has one of the highest percentages of female legislators globally. This representation has contributed to the advancement of gender equality and women's rights in the country.
2. Women's political participation in Nordic countries: Nordic countries such as Sweden, Norway and Finland have significantly promoted women's political participation through gender quotas and supportive policies. These countries have achieved high levels of female representation in parliament and government, leading to more inclusive and effective governance.[29]
3. Empowering women politicians in Kenya: Kenya's initiatives to support women's political participation include training programs, mentorship, and advocacy efforts. These programs have helped increase the number of women in political offices and improve their effectiveness as leaders.[30]

These case studies highlight the necessity of decreasing gender disparities in political participation to achieve sustainable development goals. Implementing gender quotas and overcoming barriers to political engagement can advance gender equality and support more inclusive and effective governance in societies. Promoting women's leadership is the solution to gender equality issues in sustainable development governance.

Conclusion

The intersection of gender equality and sustainable development is crucial for advancing global progress and ensuring equitable growth. In this chapter, an attempt has been made to examine various aspects of gender equality in relation to sustainable development, including the impact of gender-based policies

29 Icelandic Ministry of Social Affairs, *Equal Pay Standard: Iceland's Approach* (Reykjavik: Ministry of Social Affairs, 2020).

30 World Bank, *Economic Empowerment and Gender Equality: Case Studies* (Washington, DC: World Bank, 2021).

on social equity, economic empowerment, political participation, and health outcomes. Gender-based policies play a crucial role in reducing disparities and promoting social equity. Initiatives aimed at closing the gender pay gap, supporting women's access to education and professional development, and implementing inclusive social policies contribute to creating a more equitable society. These measures enhance women's quality of life and drive overall social progress by fostering a more inclusive and diverse environment.

Economic empowerment stands as a cornerstone of sustainable development. Gender equality in the economic sphere consists of having a discourse and finding solutions for the gender pay gap, supporting women's participation in the workforce, and promoting women's entrepreneurship. Effective policies and programs that provide access to financial services, ensure equal pay, and support women-led businesses are vital for achieving economic growth and reducing poverty. Nations can harness the full potential of their populations and drive sustainable development by removing barriers to economic participation. Increasing women's political participation and representation is crucial for ensuring diverse perspectives are included in decision-making. Gender quotas, support for women candidates, and efforts to overcome barriers to political engagement are essential for achieving gender parity in political leadership. When women are represented in political institutions, policies are more likely to cater to the needs of all citizens, leading to more inclusive and effective governance.

Reducing disparities in health outcomes, improving access to healthcare services, and doing the needful to lower gender-based violence are essential to aim for gender equality in health. Ensuring that healthcare systems are responsive to the needs of all genders, providing comprehensive maternal and reproductive health services, and finding ways to lower barriers to access care are vital components of promoting health equity. Health outcomes can be better and support the overall well-being of people. It is important to note that gender equality refers not only to equal and fair representation for people of all genders, including men, women, and non-binary individuals but also to respect any sexual orientation they may have. To sum up, achieving gender equality is integral to sustainable development. The issues and solutions discussed highlight the importance of integrating gender considerations into all development planning and implementation aspects. Advancement towards a more equitable and sustainable future can be realised through making inclusive policies. Policies and strategies to achieve gender equality benefit women and marginalised groups and contribute to broader social and economic development goals, fostering a more just and prosperous world for all.

Suggested Readings

Agarwal, B. (2010) *Gender and Green Governance: The Political Economy of Women's Presence Within and Beyond Community Forestry*. Oxford: Oxford University Press.

Arora-Jonsson, S. (2011) 'Virtue and Vulnerability: Discourses on Women, Gender, and Climate Change', *Global Environmental Change*, 21(2), pp. 744–751.

Bradshaw, S. and Fordham, M. (2015) 'Double Jeopardy: Women, Climate Change, and Disaster', *Disasters*, 39(s1), pp. s54–s75.

Dahlerup, D. (2018) *The Nordic Experience in Gender Equality*. Cambridge: Cambridge University Press.

Gender and Climate Change: Advocacy Brief (2018) Geneva: GGCA. Available at: https://www.gender-climate.org/resources (Accessed: 31 August 2024).

Global Alliance for Gender Equality in Public Life (2019) *Annual Report*. Available at: https://www.globalgenderalliance.org/annual-report (Accessed: 31 August 2024).

Global Burden of Disease Study (2020) *Global Burden of Disease Study 2019*. Available at: https://www.healthdata.org/news-release/global-burden-disease-study-2019 (Accessed: 31 August 2024).

Global Girls' Education Initiative (2019) *Annual Report*. Available at: https://www.globalgirlseducation.org/annual-report (Accessed: 31 August 2024).

Global Health Observatory (2020) *Non-Communicable Diseases*. Geneva: World Health Organization. Available at: https://www.who.int/data/gho/data/themes/topics/noncommunicable-diseases (Accessed: 31 August 2024).

Global Partnership for Education (GPE) (2020) *Global Education Monitoring Report 2020*. Washington, DC: GPE.

Government of Kenya (GoK) (2013) *National Climate Change Action Plan 2013-2017*. Nairobi: GoK.

International Labour Organization (ILO) (2019) *Women and the Future of Work*. Geneva: ILO.

International Labour Organization (ILO) (2019) *Work and Family: A Study of Employers' Practices*. Geneva: ILO.

IPU (2022) *Women in Politics Map*. Geneva: Inter-Parliamentary Union. Available at: https://www.ipu.org/resources/publications/infographics/2022/women-politics-map (Accessed: 31 August 2024).

Norton, C., McConnell, J., and Smith, K. (2019) 'Gender Differences in Mental Health: A Review of the Literature', *Journal of Mental Health*, 28(1), pp. 1–10.

OECD (2021) *Education at a Glance 2021*. Paris: OECD.

OECD (2019) *Women's Entrepreneurship: Policy Recommendations*. Paris: OECD.

OECD (2020) *Closing the Gender Pay Gap: Policy Insights*. Paris: OECD.

Oxfam (2017) *Gender-Responsive Disaster Risk Reduction: A Call for Action*. Oxford: Oxfam.

UN Women (2015) *Progress of the World's Women 2015-2016: Transforming Economies, Realizing Rights*. New York: UN Women.

UN Women (2020) *Health and Gender Equality*. Available at: https://www.unwomen.org/en/what-we-do/health-and-gender-equality (Accessed: 31 August 2024).

UN Women (2020) *Gender and Environment*. Available at: https://www.unwomen.org/en/what-we-do/environment-and-sustainability (Accessed: 31 August 2024).

UN Women (2020) *The State of Women's Entrepreneurship*. New York: UN Women.

UN Women (2021) *Women's Political Participation: Progress and Challenges*. New York: UN Women.

United Nations Development Programme (UNDP) (2012) *Bangladesh Climate Change Strategy and Action Plan*. Dhaka: UNDP.

UNDP (2019) *Gender Equality and Political Participation: A Global Overview*. New York: United Nations Development Programme.

UNDP (2019) *Gender-Responsive Economic Policies: A Global Overview*. New York: United Nations Development Programme.

UNEP (2019) *Gender and Environment*. Nairobi: United Nations Environment Programme. Available at: https://www.unep.org/explore-topics/gender (Accessed: 31 August 2024).

UNESCO (2020) *STEM Education for Girls: Best Practices and Policy Recommendations*. Paris: UNESCO.

UNESCO (2021) *Girls' Education in Rwanda: Progress and Challenges*. Paris: UNESCO.

UNFCCC (2019) *Gender and Climate Change*. Bonn: United Nations Framework Convention on Climate Change. Available at: https://unfccc.int/topics/gender (Accessed: 31 August 2024).

UNFPA (2016) *State of World Population 2016: Humanity's Future – In Our Hands*. New York: United Nations Population Fund.

UNICEF (2019) *Education for All: Global Report*. New York: UNICEF.

USAID (2017) *Women's Land Rights in Ethiopia: Achievements and Challenges*. Washington, DC: USAID.

USAID (2018) *Global Health Initiative Annual Report*. Washington, DC: USAID.

World Bank (2014) *Global Financial Inclusion Database*. Washington, DC: World Bank.

World Bank (2016) *Gender Equality and Climate Change: The Importance of Gender in Climate Policy*. Washington, DC: World Bank.

World Bank (2021) *Empowering Women in Agriculture: Opportunities for Development*. Washington, DC: World Bank.

World Bank (2017) *Financial Inclusion and Gender Equality*. Washington, DC: World Bank.

World Bank (2017) *World Development Report 2017: Education and the Future of Work*. Washington, DC: World Bank.

World Bank (2018) *Education and Gender Equality: Policy Brief*. Washington, DC: World Bank.

World Bank (2018) *Women's Economic Empowerment: Policy Brief*. Washington, DC: World Bank.

World Bank (2020) *Girls' Education in Rwanda: Achievements and Future Directions*. Washington, DC: World Bank.

World Bank (2020) *Women's Participation in the Workforce: Global Trends*. Washington, DC: World Bank.

World Health Organization (WHO) (2017) *Mental Health Action Plan 2013-2020*. Geneva: WHO.

World Health Organization (WHO) (2018) *Maternal Mortality*. Available at: https://www.who.int/news-room/fact-sheets/detail/maternal-mortality (Accessed: 31 August 2024).

World Health Organization (WHO) (2019) *Global Health Estimates: Leading Causes of Death*. Geneva: WHO.

World Health Organization (WHO) (2020) *Access to Healthcare*. Available at: https://www.who.int/health-topics/universal-health-coverage (Accessed: 31 August 2024).

World Health Organization (WHO) (2021) *Non-Communicable Diseases: Gender and Health*. Geneva: WHO.

WWF (2018) *Community-Based Conservation in Kenya: Case Studies and Successes*. Available at: https://www.worldwildlife.org/initiatives/community-based-conservation (Accessed: 31 August 2024).

Integrating Governance and Environmental Stewardship in Sustainable Development

Introduction

The term 'environment' encompasses a broad range of meanings, with divergent interpretations depending on the context. Derived from the French word *environer*, meaning to encircle or surround, the term reflects the way the environment includes all living and non-living elements in which organisms exist. The environment can be understood as the totality of conditions that shape the existence and development of life, comprising both abiotic (non-living) and biotic (living) components. In modern discourse, environmental concerns remain closely tied to the concept of sustainable development, which aims to balance economic growth with environmental protection and social equity. Integrating governance into this framework becomes essential for prioritizing environmental stewardship in development policies and practices. Effective governance ensures mechanisms are in place for societies to manage environmental resources, regulate human activities, and promote accountability and transparency in decision-making processes.

Environmental stewardship involves the responsible management and care of the environment, with an emphasis on sustainability, conservation and the restoration of ecosystems. Governments, civil society and businesses all play a critical role in ensuring the natural world is safeguarded. Governance that aligns with environmental stewardship ensures environmental concerns are considered an integral part of sustainable development rather than being overshadowed by economic priorities. This integration of governance and environmental stewardship leads to sustainable development that protects the well-being of both current and future generations, fostering a healthy environment, equitable social conditions and a thriving economy.

Environment Defined

In its broadest sense, the environment includes water, air, soil, flora and fauna. However, it can consist of natural resources, the characteristic aspects of landscape and property forming part of the cultural heritage. Some definitions also include the public services and uses that accrue from natural resources. The environment is being damaged by the rapid use of natural resources to the extent of exploitation. Industrialization has increased, polluting the atmosphere, biosphere and hydrosphere. Property forming the cultural heritage is not usually included in the environment because it is not a part of the natural resources; hence, it is considered secondary to the environment. Environment can be defined as the sum total of all surroundings of living organisms, natural forces and other living things, which provide conditions for development and growth as well as danger and damage.

Environmental stewardship and governance

The United Nations Development Programme (UNDP) defines governance as 'the exercise of economic, political, and administrative authority to manage a country's affairs at all levels. It comprises the mechanisms, processes, and institutions through which citizens and groups articulate their interests, exercise their legal rights, meet their obligations, and mediate their differences'.[1] However, governance and environmental stewardship are interconnected concepts essential for sustainable development. The US Environmental Protection Agency (EPA) defines environmental stewardship as 'the responsibility for environmental quality shared by all those whose actions affect the environment'. This includes actions aimed at protecting and conserving the environment for future generations. It emphasises the ethical and sustainable use of these resources to preserve them for future generations. Stewardship goes beyond mere compliance with environmental regulations; it requires proactive efforts to protect and enhance the natural environment through conservation, restoration and sustainable practices.[2]

Need for environmental Stewardship

Environmental stewardship is crucial for sustainable development because it ensures the responsible management and protection of natural resources, which are foundational to long-term economic and social well-being. Sustainable development aims to meet the needs of the present without compromising the ability of future generations to meet their own needs. This balance requires a careful and conscious effort to manage environmental resources that do not deplete them or cause irreversible damage. The concept of environmental stewardship involves recognizing the interconnectedness of ecosystems and human activity and the need to maintain the health and productivity of the environment. It

---

1 UNDP, *Governance for Sustainable Human Development: A UNDP Policy Document* (New York, 1997).

2 US EPA, *Environmental Stewardship: Guide Your Action to Protect the Environment* (Washington DC, 2005).

calls for practices that reduce pollution, conserve biodiversity, and use resources efficiently. Societies can avoid the detrimental impacts of environmental degradation, such as climate change, loss of biodiversity, and resource scarcity, which can undermine economic stability and social equity by integrating environmental stewardship into development strategies.

In the context of sustainable development, environmental stewardship also plays a crucial role in fostering resilience to environmental challenges. For example, protecting forests and watersheds can mitigate the impacts of natural disasters like floods and droughts while promoting sustainable agriculture and enhancing food security. Furthermore, stewardship encourages the participation of communities in the decision-making processes that affect their environment, leading to more inclusive and equitable development outcomes.[3]

Environmental damage

Environmental damage has raised concerns at all levels, including the international, national and local levels. Environmental damage can be to the environment, on the one hand, and damage to the environment, on the other. Damage to unowned natural resources affects collective rather than individual interests. Incidents that affect such collective interests do not, generally speaking, give rise to legal rights.[4] The traditional liability rules are primarily concerned with protecting individuals' interests, but in cases of damage to the environment itself, such interests are indirectly affected. The environmental damage is felt evidently as an influence on human beings. We see the rivers and lakes drying up, extreme and delayed climates, shrinking glaciers, floods and drought happening more and more in the last few decades. The environmental concern was not in the UN Charter in 1945 but has been gaining increasing attention since then. Various international organizations have been raising ecological issues with the UN. It is essential to study the evolution of environmental administration as a discipline.

Role of governance in sustainable development

Governance is pivotal in sustainable development, as it establishes frameworks, policies and systems that align economic, social and environmental goals. Effective governance ensures that development initiatives are inclusive, equitable, and environmentally sound, balancing the needs of present and

3 WCED, *Our Common Future* (Oxford, 1987).

4 H. Bocken, 'Developments with respect to compensation for damage caused by pollution', in *The Gradual Convergence*, ed. by B. S. Markesinis (Oxford, 1994).

future generations. It provides the institutional structures necessary for enforcing laws, setting standards, and monitoring progress towards sustainability targets. Governance promotes accountability and empowers communities to participate in the development process through transparent decision-making and stakeholder engagement. This participatory approach helps address various regions' diverse needs and challenges, ensuring that development is sustainable and resilient to environmental and social changes.

Good governance fosters innovation by creating environments where sustainable practices and technologies can thrive. It supports the integration of environmental stewardship into policy-making, ensuring that the exploitation of natural resources is minimised and that the ecological balance is maintained. Governance also plays a critical role in managing conflicts that arise from competing interests, such as economic growth versus environmental conservation, by facilitating dialogue and consensus-building among stakeholders.[5]

The origins of environmental administration

The following is a timeline that briefly describes the evolution of environmental administration.

1864 Marsh wrote a book, *Man and Nature*, or *Physical Geography as Modified by Human Action*, after Darwin's theories, and cautioned that 'man is everywhere a disturbing agent. Wherever he plants his foot, the harmonies of nature are turned to discords'. He advocated prudent policies in forestry practices.

1870 Agassiz, a French scholar, gave his views on natural history, although he did not accept Darwin's theories.

1873 Mark Twain and Charles Dudley Warner published *The Gilded Age: A Tale of Today*.

1901–1909 Theodore Roosevelt, as the President of the USA, did a lot for environmental protection by setting aside lands for national parks and museums. By the mid-nineteenth century, environmentalists came to the forefront.

1935 Wilderness Society was formed to protect the wilderness.

5 United Nations, *World Public Sector Report 2018: Working Together - Integration, Institutions, and the Sustainable Development Goals* (New York, 2018).

1914 International concern was voiced by private people to protect wildlife, marine animals and fisheries in 1872. A consultative commission at Berne was established for the international protection of nature. It maintained a legal status between the two world wars as the first intergovernmental agency for environmental protection.

1947 The Swiss League sponsored a conference at Brunnen for the protection of nature attended by twenty-four countries and nine international organizations, which adopted a draft constitution for establishing the International Union for Conservation of Nature, followed by the UNESCO Conference.

1950s Concern over the balance between the environment and human life took the international stage.

1960s Environmental concerns were voiced forcefully in the United States, but the communist world continued destroying the environment in the name of the environment.

1970s The focus was on the biophysical environment, which laid the foundation for modern environmentalism.

1972 The Stockholm Conference, more popularly known as the 'Stockholm spirit of compromise', occurred following severe damage to thousands of Sweden's lakes from acid rainfall due to severe pollution in Western Europe.

1972 An intergovernmental conference was held in London to form a worldwide regulation of the problem.

1973 Convention on International Trade in Endangered Species of Wild Flora and Fauna (CITES).

1976 The FAO held a convention in Barcelona to protect the marine environment against Mediterranean Sea pollution. It covered three aspects of marine pollution, including discharge of oil, ocean dumping and land-based pollution.

1982 The Nairobi Declaration was adopted in Nairobi to celebrate the tenth anniversary of the Stockholm Conference on Human Environment in 1972.

1985 The Vienna Convention was adopted on 22 March by the Conference of Plenipotentiaries organised by the UNEP.

1987 The Montreal Protocol on substances that deplete the ozone layer came into force in 1987 and was finally adopted in 2000 by 173 States as a party to it.

1989 The Helsinki Declaration was made on protecting the ozone layer to phase out CFC production and consumption by 2000.

1989 Convention on Transboundary Movement of Hazardous Wastes, enacted in 1992.

1992: The Earth Summit was held, and over 150 countries attended it.

1992 The UN Framework Convention on Climate Change (UNFCCC) was held to stabilise greenhouse emissions at levels that would prevent dangerous anthropogenic interference in the global climate.

1992 The Convention on Biological Diversity was a legally binding framework treaty ratified by 180 countries.

1992 The Rio Declaration on Environment and Development produced a document at the UN Conference on Environment and Development (UNCED), popularly known as the Rio Earth Summit. It gave twenty-seven principles to guide future sustainable development in the world.

2012 Rio Summit celebrated its twentieth anniversary by the Rio + 20 UN Conference on SD.

2015 Environment and development are vital in SDG Agenda 2030. The laws and principles governing the environment, such as the 'precautionary principle', are an important move forward. Besides the programmes mentioned above and the international response to environmental issues, hundreds of bilateral and multilateral environmental treaties create States' rights and obligations. The UN Environment Programme (UNEP) and the UN Commission on Sustainable Development has negotiated many of these treaties.

The complimentary dynamics of environment and development

The UNGA gave an urgent call to formulate a global agenda for change' in which Norway Prime Minister Gro Harlem Brundtland took the lead, also popularly known as 'the mother of sustainable development'. A report, 'One Common Future', was published by the World Commission on Environment and Development. Instead of pondering over the ever-increasing environmental decay, poverty and hardship in an ever more polluted world among ever-increasing resources, the report focused on the possibility of a new era of economic growth based on policies that sustain and expand the environmental resource base.[6] However, the complementary dynamics of environment and development need some retrospective study of how our planet reached this situation. The words' environment' and 'development' have become the so-called 'buzzwords' of the intellectuals discussing the fate of 'our' planet at all levels of forums. The prefix 'our' can mean ownership and emphasise man's vocation to master and run this common property, or it can imply participation and highlight man's dependence on an encompassing reality.

Wolfgang Sachs is known as one of the pioneers in building development administration for his insight into how the new fad of development and developmental aid of the West to the Southern Hemisphere. On the other hand, the Brundtland Report wedded the long-awaited vision of growth. The notion of industrialization linking growth and squandering national resources was difficult to shake off. The 1980s was the period of the emergence and strengthening of the NGOs, primarily in environmental protection.

The pretension of development

Throughout the 1970s, environmentalism was considered as opposed to reducing poverty. The development ideology of the West claims to exist under the pretence of development. Environmental degradation has been the product of industrialization and is unrelated to poverty, but this soon changed as deforestation and desertification increased worldwide. People experiencing poverty were designated as the agents of destruction, which led to widespread campaigns to promote 'environmental consciousness', ignoring the fact that they were the victims of the environmental degradation caused by the industry big wigs in the name of economic growth.[7] The relationship between environment and development is explained in Figure 30.

1. Trade-off: The more industrialization in a society, the more natural resources are used while polluting land, air, water and soil, resulting in environmental damage. This dilemma is more prevalent in developing countries as they have to ramp up industrial progress to compete in the world markets owing to globalization. They fail to maintain the balance

6 *Development Dictionary: A Guide to Knowledge as Power*, ed. by W. Sachs (London, 2010).
7 Adapted from Sachs, *Development Dictionary*.

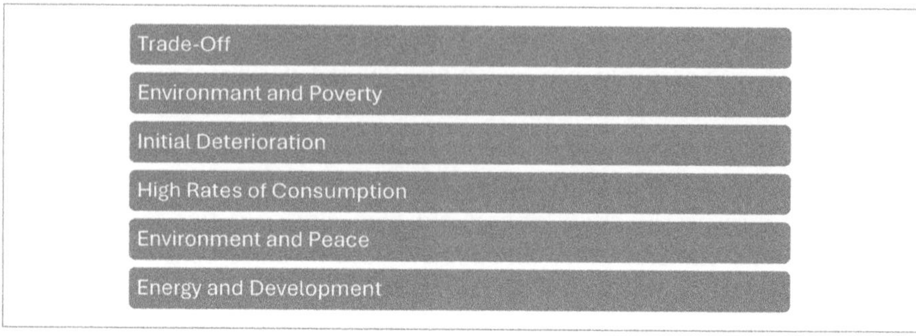

Figure 30 Environment and development

between their economic growth and ecological balance. Adding to this imbalance is the exponential rate at which the population is increasing in the developing nations.

2. Environment and poverty: The Southern Hemisphere has relatively higher poverty rates, and these countries are developing. Since they also need to improve their socio-economic indicators, environmental degradation due to industrialization will continue to add to the impending ecological collapse. Many land and water animals have become extinct or are on the way to vanishing from our planet. However, many environmental and development studies have revealed that living standards concerning water, hygiene and life expectancy increase with development.

   One of the primary barriers to integrating environmental stewardship into sustainable development is economic constraints. Developing countries, in particular, face significant financial limitations that restrict their ability to invest in sustainable practices. The costs associated with implementing environmental regulations, adopting green technologies and transitioning to sustainable economic models can be prohibitive. Furthermore, there is often a lack of access to financial resources, including international funding and investment, to support these initiatives. This economic barrier is compounded by the short-term focus of many governments and businesses, which prioritise immediate economic gains over long-term environmental sustainability.

3. Initial deterioration: Development initially deteriorates the environmental indicators because of the amounts of sulphur dioxide produced by the manufacturing industry. However, alternate energy consumption methods are introduced instead of biomass fuel as development occurs. Research studies have shown that higher technology and services have reduced pollution per dollar of GDP growth.

4. High consumption rates: As nations develop, the carbon dioxide emissions also increase. Some developed countries have managed to handle waste due to increased resources. However, by and large, industrial waste due to increased consumption has adversely affected the environment. One cannot retard the growth of development in developing countries in the name of environmental changes. Still, technical advancements in developed countries can help them to deal with environmental issues better. Development aid must come in ways to help them, especially in reducing the adverse effects.
5. Environment and peace: Peace and security issues have been explored extensively; however, they can be found in the principles and goals laid out by the Rio Declaration in Principles 25 and 26. Serious efforts must be made to peacefully resolve disputes emerging due to natural resources, transboundary environmental damage, or global commons. Moreover, a nation torn with civil war, strife and violence cannot pay any attention to environmental concerns as its whole concentration is to survive. According to the UN, security does not form the absence of poverty, internal conflict, mass destruction weapons, supernational organised crime and the like. Environmental security emerged from the understanding that environmental degradation or scarcity of natural resources can cause and result from political or military conflict. Wars over natural resources are a significant concern, so appropriate use must be made.
6. Energy and development: Underdeveloped countries and developing nations usually use biomass fuel, which leads to deforestation, soil degradation and desertification. The energy-development-environment connection is a vital synergy to be studied at each level. Initiatives are being taken to reduce the use of biomass fuel. For example, India is making significant efforts in this area of renewable energy. Awareness of solar energy has made people use it more with incentives given by the government. A part of this campaign is building green housing societies designed to use renewable energies that are naturally cool in the summer and warm in the winter, reducing power use to maintain temperatures. Energy-efficient electric appliances are being manufactured to save energy from 5 to 15 per cent. Such innovative steps can lead to energy conservation, and development can be sustained without extreme environmental damage, considering the costs of capital equipment and labour requirements.
7. Political and institutional barriers: Political will and institutional capacity are crucial for successfully integrating environmental stewardship into sustainable development, but these are often lacking. Governments may face political pressure to prioritise economic growth over environmental

protection, particularly in regions where economic development is essential for poverty alleviation. Institutional barriers, such as weak governance structures, inadequate enforcement of environmental laws, and a lack of coordination between different government agencies, can impede the implementation of sustainable development policies. The absence of clear regulatory frameworks and the fragmentation of responsibilities among various institutions further complicate the integration process.

8. Social and cultural challenges: Social and cultural factors also significantly hinder the integration of environmental stewardship into sustainable development. Public awareness and education about environmental issues are often limited, leading to a lack of understanding and support for sustainable practices. Cultural values and traditions may prioritise economic development and resource exploitation over environmental conservation, making shifting towards more sustainable practices difficult. Moreover, in many communities, particularly in rural areas, there is resistance to change due to fears of economic displacement or loss of livelihoods, which can result from adopting environmentally sustainable practices.[8]
9. Technological and infrastructural barriers: The lack of access to appropriate technology and infrastructure is another significant challenge in integrating environmental stewardship into sustainable development. Many developing countries do not have the technological capacity to implement sustainable practices, such as renewable energy systems, waste management facilities, or efficient water management techniques. The infrastructure required to support these technologies is often underdeveloped or non-existent, further hindering progress. Additionally, technology transfer from developed to developing countries is usually limited by intellectual property rights, high costs, and a lack of technical expertise.
10. Global and transboundary issues: Environmental problems, such as climate change, deforestation, and pollution, do not respect national borders and require coordinated international efforts to address them. Conversely, global governance structures for environmental issues are often weak, and effective international cooperation and enforcement mechanisms are lacking. The disparities in environmental responsibilities and capabilities between developed and developing countries create tensions and hinder collective action.[9]

---

8 J.D. Sachs, *The Age of Sustainable Development* (New York, 2015).

9 United Nations, *Transforming Our World: The 2030 Agenda for Sustainable Development* (New York, 2015).

To sum up, the environment and development have an essential relationship, mostly labelled as inverse, but can be changed by taking precautionary steps. One cannot ignore the development needed to save the planet as it is a sustained elevation of society and the social system geared towards a better or more humane life. Integrating environmental stewardship into sustainable development is fraught with challenges and barriers that require comprehensive and coordinated efforts to overcome. These obstacles necessitate a multifaceted approach that includes economic support, political commitment, institutional strengthening, public education, and technological innovation. Global cooperation and the sharing of responsibilities among nations are essential to achieve sustainable development that is genuinely inclusive and environmentally sound.

Integration of environmental stewardship in governance

Environmental stewardship is critical in ensuring sustainable development, where economic growth is achieved without compromising the ability of future generations to meet their needs. Integrating environmental stewardship into governance involves embedding environmental considerations into the decision-making processes at all government and private sector levels. This integration requires a holistic approach where environmental policies are not isolated but are part of a broader governance framework that includes economic and social policies. Effective environmental stewardship in governance involves the establishment of clear regulations, the enforcement of environmental laws, and the promotion of sustainable practices across all sectors.

Governments must engage with various stakeholders, including businesses, civil society and indigenous communities, to ensure that environmental policies are inclusive and reflect diverse perspectives. Integrating environmental stewardship into governance entails adopting innovative approaches, such as using technology to monitor environmental impact and the development of market-based instruments like carbon pricing to incentivise sustainable practices. Transparent and accountable governance structures are essential for successfully integrating environmental stewardship. This includes precise reporting mechanisms, independent oversight bodies, and the involvement of the public in environmental decision-making. By integrating environmental stewardship into governance, governments can ensure that economic development is sustainable, socially inclusive, and environmentally sound.[10]

10 D. H. Meadows et al., *Limits to Growth: The 30-Year Update* (White River Junction, VT, 2004).

Sustainable livelihoods

The sustainable livelihoods frameworks seek a more comprehensive and integrated approach to poverty than the traditional interpretations, which essentially consider poverty as a narrow set of indicators such as income and productivity.[11] The concept of sustainable livelihood attempts to go beyond the conventional definitions and approaches to poverty eradication. These were too narrow because they focused only on certain aspects or manifestations of poverty, such as vulnerability and social exclusion. The Brundtland Commission first introduced the idea of sustainable livelihoods on environment and development to link socio-economic and ecological considerations in a cohesive, policy-relevant structure.[12] The 1992 conference on environment and development advocated for achieving sustainable livelihood as a broad goal for poverty eradication. It stated that sustainable livelihoods could serve as 'an integrating factor that allows policies to address 'development, sustainable source management and poverty eradication simultaneously'.[13]

Framework for sustainable rural livelihoods according to the Institute for Development Studies, UK

The Institute for Development Studies (IDS) in the United Kingdom gave a framework regarding livelihood resources conceptualised as different types of capital to stress their role as a resource base. Four types of capital are identified in the IDS framework, which are given in Figure 31.

1. Natural capital: The natural resource stocks (soil, water, air, genetic resources) and environmental services (hydrological cycle, pollution sinks and so on) from which the resource flows and services valid for livelihoods are derived.
2. Economic or financial capital: The capital base (cash, credit, savings and other economic assets, including basic infrastructure and production equipment and technologies) essential for the pursuit of livelihood strategy.
3. Human capital: The skills, knowledge, ability of labour, good health and physical capability are crucial for successfully pursuing different livelihood strategies.

11 <https://atha.se>, accessed 13 August 2024.

12 < https://www.sida.se>, accessed 13 August 2024.

13 UNDP, 'Promoting sustainable livelihoods: A briefing note submitted to the executive committee' (New York, 1997).

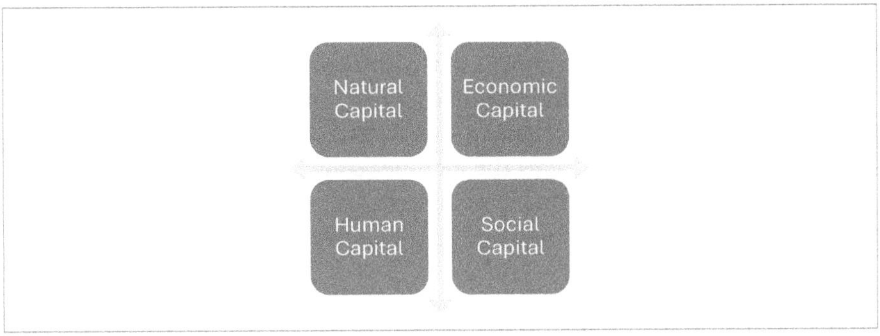

Figure 31 Types of capital

4. Social capital: The social resources (networks, social claims, social relations, affiliations, associations) upon which people draw when pursuing different livelihood strategies requiring coordinated actions.

The sustainable livelihood approach and its principles

The sustainable livelihoods approach is a way of thinking about development activities' objectives, scope and priorities. It is based on evolved thinking about how the poor and vulnerable live and the importance of policies and institutions. It helps formulate development activities that are as follows:

- People-centred
- Responsive and participatory
- Multi-level
- Conducted in partnership with the public and private sector
- Dynamic
- Sustainable

A sustainable livelihood approach does not replace participatory or integrated rural development but connects people with the overall environment that influences the outcomes of livelihood strategies. The ADB proposed appreciative inquiry as a tool for industry to avoid negative approaches to problem-solving. It is a highly inclusive process that maximises the positive in which a community takes responsibility for generating and gathering information and then forms strategies based on the most positive experiences of the past.[14]

14 <https://www.un-documents.net/ocf-02.htm>, accessed 13 August 2024

Paris Agreement

The biggest challenge of the present day is climate change. The latest initiative has been the Paris Agreement within the UNFCC, dealing with GHG emissions mitigation, adaptation and finance, which started in 2020. The Paris Agreement was signed in 2015 and requires the member countries to make binding commitments to curb carbon dioxide emissions to keep global average temperatures from rising above 1.5°C as compared to pre-industrial years.

Environmental law: The transboundary perspective

Environmental laws are the standards that governments establish to manage natural resources and environmental quality which include pollution of air, water, forests and wildlife, hazardous waste, agricultural practices, wetlands and land use planning. Each country's government, as well as the state and local institutions, adopts its laws. National laws can impact other nations through their products or transboundary pollution issues in common borders. In such cases, only extraterritorial applications of national law are applied.

However, international law concerns agreements among nations, citizens, or corps of different nations. Such agreements or treaties among other countries are called public international law but contracts between private parties (corps or citizens) living in separate countries are called 'private international law'.

International Environmental Law (IEL) is divided into the following two types:

1. Soft international law: It consists of agreements or principles that aim to influence individual nations to consider specified norms with respect and include them in the national law. Such norms are not enforceable, but they lay down principles and standards to maintain the sovereignty of nations. IEL comes under the category of soft law as Article 38 of the ICJ. The non-legally binding normative agreements make it easier to press dissenters into conforming behaviour as States are free to use political pressure to influence others to alter their policies. However, they cannot demand that others conform to similar legal norms. They represent the international community's will to resolve a pressing global problem over obligations.
2. Hard international law: Treaties and customary international laws are hard law. 'Hard international law' refers to legal instruments with binding force in international law, such as treaties and customary international law. Treaties are formal agreements between states that are legally binding upon the parties that sign and ratify them. These agreements cover a wide range of issues, from human rights to environmental protection and

once ratified, they impose obligations on the participating states under international law. Customary international law, in contrast, is derived from consistent and general practice by states, accompanied by a belief that such practice is legally required. This law evolves and is considered binding on all states, regardless of whether they have explicitly consented to it. Customary law plays a crucial role in areas where treaties may not exist, filling gaps and providing a framework for international relations.[15]

International environmental law

It is interdisciplinary, simultaneously intersecting and overlapping with numerous other research areas, including political science, economics, human rights and navigation admiralty. The trend in International Environmental Law began with the 1972 Stockholm Declaration of the UN Conference on the Human Environment, followed by national laws to preserve the environment. The Rio Declaration also falls into this category.

Common principles in the IEL

Some common principles and concepts have found a place in the IEL. They are as follows:

1. Principle of responsibility and liability for environmental damage: Whether the damage is domestic or transboundary, civil or public, corporates not only have the duty to compensate but also must prevent such damage. This principle has found a place in all IELs.
2. Intergenerational equity: The principle of intergenerational equity, often highlighted in international environmental law (IEL), such as in the Stockholm Declaration, Rio Declaration, UNEP, and CSD, suggests that current generations must manage the environment in a way that ensures its availability and health for future generations. Although these instruments widely acknowledge this principle, its legal value and binding force remain underdeveloped and contested. While it may not yet carry the weight of enforceable law, the principle extends beyond mere aspiration, reflecting a growing normative standard that influences the development of international environmental policies and legal frameworks.
3. Human rights: The expectation for human rights in international environmental agreements like the Stockholm Declaration, the Rio Declaration and the Commission on Sustainable Development (CSD) is significant.

15 M. N. Shaw, *International Law*, 8th edn (Cambridge, 2017), 12–15

These declarations incorporate the language of human rights, recognizing the essential connection between environmental protection and human well-being. However, the challenge lies in the practical implementation of these principles. Despite the strong rhetorical commitments, the actual application and enforcement of human rights within the context of these agreements remain an ongoing issue, with many questions about how to achieve effective integration and accountability still unresolved.

4. Development and environment: Development and environment have been interlinked ever since development started as an organised strategy and a process. There is no legal binding based on this principle, but many natural laws have included the issue of the environment in all development projects, leading to sustainable development.

Some other common features in the large body of agreements are:

- Sovereignty of states over their natural resources and the responsibility not to cause environmental damage
- Cooperation between the states and being good neighbours
- The precautionary principle that the responsibility for harmful actions on the environment falls on the person taking the action
- Polluter pays
- Sustainable development

The legal consequences in each case depend on the facts and circumstances of that particular case. The onus of making these principles into binding laws falls on the ICJ. Thus, International Environmental Law (IEL) has developed robust principles and concepts that reflect the global community's commitment to addressing environmental challenges. However, transforming these principles into binding international laws largely depends on the International Court of Justice (ICJ) and the collective will of the global community. As the world continues to confront environmental issues, these principles' ongoing development and enforcement will be crucial in achieving sustainable development and protecting the global environment for future generations.

Development implications of international environment agreements and laws

There is an increasing concern over the economic implications of signing up to environmental agreements. Developing countries feel threatened concerning economic competitiveness in the light of stringent standards. Also, they must bear the additional cost to follow such protective measures without compensation. Lack of funds makes the developing nations forgo environmental protection to increase economic growth. Realizing this issue, some agreements like the Montreal Protocol 1987 on Substances that Deplete the Ozone Layer have included compensation in Article 10. It considers the needs of developing countries to promote financial and technical assistance, which facilitates the compliance of signatories with the Protocol requirements. Principle 12 of the Stockholm Declaration (1972) highlights the need to consider the costs that developing countries might face when implementing environmental protections. It also suggests that financial assistance should help these countries manage those costs.

Mechanisms of the IEL: The Kyoto Protocol Agreement

One hundred sixty countries have signed and ratified the Kyoto Protocol by 2006. The countries who ratified the Kyoto Protocol are legally bound to uphold it. Its main goal is to cut global emissions of GHGs, such as CO2, methane, nitrogen oxide, sulphur hexafluoride, HFCs, and PFCs. The Kyoto Protocol aims to stabilise GHG concentrations in the atmosphere at a level that would prevent dangerous anthropogenic interference with the climate system. The Treaty agreed on the following three basic premises:

- The largest share of global GHG emissions is from developed countries.
- Emissions per capita in developing countries are still reasonably low.
- The share of global emissions in developing countries will grow to meet social and development needs.

The Kyoto Protocol emissions trading

Some countries that committed to a fixed level of emissions and ratified the agreement set their targets higher than their current production levels. Emissions trading was a mechanism of selling extra credits to countries exceeding their limit on the open market. Emission trading rewards countries that meet their target emission level and provides financial incentives to meet or be below levels.

Ways to earn credits

The Kyoto Protocol gave three ways to earn and trade emissions, which are as follows:

- The Clean Development Mechanism programme: They can earn credits by investing in projects in developing countries.
- Joint implementation: They can invest in projects to help other countries reach their emission targets set out by the Protocol.
- International emissions trading: Buying and selling emission credits.

However, the Kyoto Protocol was criticised because the developed countries must pay and supply emission-reducing technology to other countries for climate-related studies and projects. Some nations who are sceptical of global warming view the Protocol as a scheme to retard the growth of developed countries and to transfer wealth to underdeveloped countries. India and China were exempted from this Treaty, for which much criticism was charged.

Environmental ethics and development

Environmental ethics is the philosophical discipline that considers the moral and ethical relationship of human beings to the environment. There is no embargo on development, but humankind has a moral obligation to preserve and conserve the non-human world. It is the ethical relationship between humans and the natural environment. Ancient India successfully maintained a sustainable balance between itself and the non-human. It was understood then that if we pressurise nature and 'Mother Earth', nature will become destructive. Presently, the earth is absorbing the human attack of pollution, waste material, deforestation, desertification, ocean and marine pollution, and so on because there is a single-word answer: development. The answer to this quandary is to aim for sustainable development.

Glimpses of ethical values can still be seen in rural India, which works in small ways to conserve the environment. For example, women in rural areas cook and keep food for their cows, dogs and birds before they themselves eat, whereas such values are ignored today. However, the issue is that in the name of economic development and progress, the environment has a place in agreements, treaties, and legislation. Still, it does not get any ethical value that needs to give the environment its due as environmental degradation worsens. These values hold more relevance in the present scenario, but Indian tradition and thought have not been brought to the international forums. At the core of environmental ethics is the idea that the environment has intrinsic value independent of its utility to human beings. This perspective challenges the anthropocentric view, which places human needs and interests above all else.[16] Environmental ethicists argue that nature should be respected and protected for its own sake, not merely as a resource for human consumption. This approach advocates for the moral consideration of all living beings and promotes the idea that humans are responsible for preserving the environment for future generations. Several ethical frameworks have been proposed within environmental ethics.

---

16 R. Attfield, *Environmental Ethics: An Overview for the Twenty-First Century* (Cambridge, 2014).

For example, deep ecology emphasises the interconnectedness of all life forms and advocates for a radical shift in human consciousness to live in harmony with nature.[17] Meanwhile, ecofeminism links environmental degradation to the exploitation of women and other marginalised groups, arguing for a more inclusive and just approach to environmental stewardship. In practice, environmental ethics informs policies and actions related to conservation, sustainability, and environmental justice. It encourages individuals and societies to reflect on their environmental responsibilities and make ethical choices contributing to the planet's well-being.

Approaches to environmental ethics

Environmental ethics emerged as a discourse and a discipline catching the attention of environmentalists globally. It gained importance in 1970 with Earth Day when environmentalists urged the philosophers to inspect the philosophical issues connected humans to their environment.

Three main theories/approaches to environmental ethics are:

1. Anthropocentrism: Anthropocentrism is the view that all environmental responsibility is derived from human interests alone. The assumption is that only humans are morally significant and have direct moral standing. Since the environment is crucial to human well-being and human survival, we have an indirect duty towards the environment, that is, a duty derived from human environments. We must ensure that the Earth remains environmentally hospitable to support human life and a pleasant place for humans to live. Epicurus, an ancient Greek philosopher, gave the principle of 'utility' based on utilitarianism and stated that to understand nature, the following must be followed:
   - Limitation of desires
   - The practice of virtues
   - The pursuit of happiness through friendship

   However, the Anthropocentrism approach was criticised because it was inadequate for conducting and regulating the environment and human relations.[18]
2. Biocentrism: It refers to all the forms of life having an inherent right to exist. This approach emerged in the 1970s. This approach was pioneered

17 H. Rolston, *Environmental Ethics: Duties to and Values in the Natural World* (Philadelphia, 1988).

18 *Global Climate Change, Environment and Energy–Global Challenges and Opportunities to Global Stability*, ed. by F. Katman (Newcastle upon Tyne, 2014).

by the Nobel Prize winner, Dr. Albert Scheitzer and later by Paul Taylor. The non-anthropocentric approach comprises four premises:

- Humans are the members of earth's community of life on the same terms of life on the same terms as all living beings.
- All species, including humans, are a part of a system of interdependence.
- The technological centre of life – the belief that all living things pursue their good in their way.
- Humans are understood as not being inherently superior to other living things.

The biocentrism approach states that there is an inherent right to live, to have freedom and not be exposed to torture and violence of all living beings, but it does not cover non-living things. However, this approach has a grave contradiction as it includes animals in its sphere which are used as food.

3. Ecocentric ethical approach: The ecocentric ethical approach proposes that humans are not the only ones in the sphere of ethics. Other living and non-living things (the ecosystem) should also be included. The environment deserves direct moral consideration and not consideration that is merely derived from human or animal interests. The environment, not just the living organisms, has moral worth. The ecological viewpoint and the conception of the integrity of nature form the basis of this approach. Many other philosophers and environmentalists have worked on the concept of environmental ethics. It is an inescapable reality of our life today which must be valued and respected by all.

Conclusion

Wolfgang Sachs is one of the pioneers in promoting sustainable development. He was sceptical about what was declared in the Brundtland Report. He argued that those affected by poverty had now been recognised within the professional consensus, meaning that there was agreement among experts on their situation. The old solution, economic growth, was once again proposed to address this new crisis, believing that growth would alleviate poverty and that environmental protection could only be achieved through a renewed focus on economic expansion. The Brundtland Report professed, 'No development without sustainability; no sustainability without development' and development emerges again, but development still eludes most undeveloped nations. Every time development does not happen, a new twist is given, such as equitable or sustainable development. Politicians and technocrats gain from the whole concept and steer the economic growth that profits them.

Ecology is an amalgamation of computer modelling, political action, scientific discipline and an all-embracing worldview. However, there is a continuous tiff between the environmentalists and the scientists who are called upon to ratify their protests. The science of ecology leads to a rise in scientific anti-modernism. Sachs points out that all international efforts move towards debating and controlling carbon dioxide emissions or scientists send satellites into space to look at our planet and its condition. All this requires extended self-management but ignores intelligent self-limitation. Instead of concentrating on whether nature exploitation needs to be maximised or minimised, attention must be paid to its optimization. The urgent call to control environmental exploitation implies that the developed countries continue to grow their industrial economy and let the others manage the environment.

It is essential to have slick mechanisms for surveillance and regulation in the mammoth task of maintaining a balance between industrial growth and nature. The state needs to be more vital in building ecological societies with less government and less professional dominance. There is no thought of reducing the root cause of industrialization and planning to develop ecological societies with less government and less professional dominance. Too much stress is laid on managerial strategies for resource efficiency and risk management. Public values, morals and ethics shape how society determines appropriate limits for production and consumption. Societies and democracies all over the globe compete with each other to achieve higher economic growth. In the words of Wolfgang Sachs, 'The ecocratic perception remains blind to diversity outside the economic society of the West'.[19]

Environment has a vital role to play in the development of economics. Environmental consciousness has been a part of our culture all over the world since ancient times and is not a recent development. The SDG Agenda 2030 has been built around the concept of sustainability in all its goals. Innovations in science and technology are being made based on sustainability and conserving the environment. Appliances are being designed to be energy efficient. Renewable energy is an upcoming area in which India is fast catching up with the leading countries in the world. However, the attitudes and behaviours among the common masses, as well as the leaders must be moulded positively towards the environment by obtaining a deeper understanding. OECD has introduced the concept of behavioural insight to help the policy-makers contribute to environmental problems and design and implement more effective policy interventions.

---

19 Sachs, *Development Dictionary*.

It encourages the application of behavioural insights to promote more sustainable consumption investment and compliance decisions by individuals and forms.[20]

Green growth and development must go hand-in-hand by employing a new approach to have economic growth to ensure services to support sustainable development. Developing countries are particularly vulnerable to climate change, environmentally, ecologically and socially. The governments must work towards the integrated development cooperation through collaboration between development and environment communities.

20 <https://www.oecd.org>, accessed 13 August 2024.

An Ethical Deficit in Sustainable Governance

The quest for sustainable development, defined by the balance between economic growth, environmental protection and social equity, is intrinsically linked to the ethical integrity of governance systems. This chapter delves into the 'ethical deficit in governance' concept and its profound impact on achieving sustainable development goals (SDGs). An ethical deficit occurs when governance structures fail to uphold transparency, accountability and fairness principles, leading to corruption, discrimination and inefficiency. These shortcomings erode public trust and hinder the effective implementation of policies designed to foster sustainable growth. This chapter highlights ethical governance's critical role in creating a just, equitable and sustainable future by exploring the various dimensions of ethical governance, ranging from corruption and accountability to environmental stewardship and human rights. Through this lens, we will examine the challenges and propose pathways to strengthen governance frameworks, ensuring they align with the ethical imperatives necessary for sustainable development.

Introduction

The concept of good governance, having no defined boundaries, extends itself in all directions. It covers the fundamental rights and duties which a government must cater to. Various world organizations have identified the indicators used to understand and analyze the level of governance being categorised as good. The most commonly used are human rights, rule of law, accountability, transparency, responsiveness, equitable, inclusive, efficient, effective and participatory. Governance is the process by which governments make decisions and execute them. Usually, it acts as a mechanism that measures how much various governments could succeed in delivering in the areas mentioned above. Global organizations conduct comprehensive analyses of economies on an international scale, focusing on governance related to markets, public-private partnerships (PPPs), political representatives and the executive, among other critical issues.

Governance is not a new concept. We, for sure, know that it existed in India in the fourth century when Kautilya wrote his famous work, Arthashastra. In contrast, a newfound interest is now seen in the study of good governance. This fact is deduced from the simple fact that although governance, no doubt, has been there ever since, how good it was must be evaluated. There is a definite difference between the developed and the developing countries in their delivery system and governance. The causes of this difference must be analyzed by world organizations such as the World Bank, International Monetary Fund and Asian Development Bank. These organizations help improve their member countries' good governance indicator ranks by adopting new policies to have better economies and regulatory structures. Good governance involves implementing the government's development plans and policies and restructuring organizations to ensure effective management systems. The eventual ultimate stakeholder, the citizen, must have order, justice, equity and accountability.

Sustainable development governance, good governance and ethics

The word 'ethics' is derived from the Greek word 'ethos', which means 'custom', denoting how people should act for the greater good normatively. It is commonly confused with 'morals', which imply the practice of ethics. In the West, the ancient Greek philosophers such as the Sophists, Socrates, Plato, Aristotle, Epicurus and Stoics wrote extensively about ethics. In medieval times, English positivists continued the school of thought in Europe. Socrates has been considered the 'Father of Ethical Philosophy' in the West, and Hobbes the 'Father of Modern Ethics'. Utilitarian and Kantian ethics propagated the concept of ethics. In India, 'dharma' refers to moral and social norms and can be related to ethics and morality. Our ancient scriptures, which date back to 5,000 BC, teach us how to live by doing the right action. All ancient scriptures of India, such as Ramayana, Mahabharata, Vedas, the Puranas and the Upanishads, tell us the importance of ethical conduct. However, the 'dharma' approach can mean morality, justice, and intrinsic quality.

Ethics forms the basis of all human behaviour, which becomes the reason for different outcomes. So, an organization must work based on ethics in every perspective because it would result in sustainable growth over a long period. There would be a considerable decrease in cost because ethics would form a robust governance system. Such an organization would finally spend less by saving on unlawful practices. Since the conception of human societies, corruption has been present all around. There is evidence of corruption in various literary works such as Dante's, Kautilya's Arthashastra and Shakespeare's plays. One of the primary reasons for its being highlighted is the role of the fourth estate, the media. Globalization has closed the gap between economies and highlighted that corruption impeded economic growth. NGOs have also been an essential factor in increasing awareness towards corruption. Since corruption retards progress in economic governance and efficiency, more attention is being paid to this problem. Lastly, the US initiated the subject and tried to resolve it by establishing international institutions as it lost out on foreign deals in the global market because of the demand for bribes from government officials.

Sustainable development governance, good governance and ethics are deeply intertwined, forming a foundational triad that underpins efforts to achieve sustainable development goals (SDGs) fairly, just and responsibly. Ethics and good governance are inherently connected, as good governance is rooted in ethical principles such as integrity, fairness and justice. Ethical governance demands that those in power act transparently, accountably, and in the public's best interest fostering trust between governments and citizens. Moreover, ethics in decision-making ensures that policies reflect values like respect for human dignity, social justice, and the common good, preventing inequality and corruption.

In ethics and sustainable development governance, intergenerational equity is a core ethical concern, emphasizing the moral responsibility to ensure future generations inherit a world without resources. This drives sustainable management practices and environmental protection efforts. Ethical stewardship compels humans to care for the planet, guiding policies that promote sustainability and conservation. Justice and equity are also central, ensuring that development benefits are distributed fairly and the most vulnerable are protected from adverse impacts.

Finally, integrating ethics, good governance and sustainable development governance requires ethical leadership that prioritises the well-being of both people and the planet over short-term gains. Ethical governance extends to corporate responsibility, with businesses expected to operate ethically, contributing to social and environmental well-being. Policies must be made with ethical considerations, ensuring they do not favour the powerful at the expense of the vulnerable and that they align with sustainability goals while upholding justice, equity, and human dignity.

Approaches to ethics

Generally, philosophers divide the ethical theories into three, depicted in Figure 32.

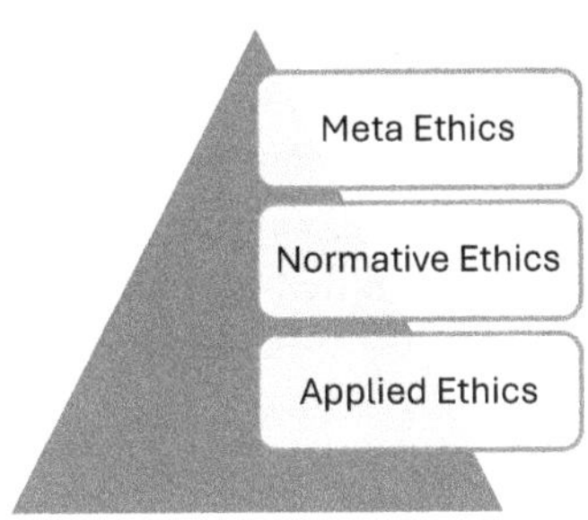

Figure 32 Approaches to ethics

- Metaethics: It deals with the nature of moral judgement, which comprises the origins and meaning of ethical principles.
- Normative ethics: It refers to the content of moral judgements and the criteria for what is right or wrong.
- Applied ethics: It deals with controversial topics such as war and animal rights.

Ethical governance

Governance and ethics have deeply connected dynamics, as governance without ethics is meaningless and ineffective, resulting in unethical and corrupt behaviour and practices. The delivery of public services by the government will be unsuccessful in reaching the citizens. Ethical governance can be described as the right and fair conduct of governmental activities to serve the greater public interest. The exercise of management activities of the government's social and economic resources that are acceptable to all stakeholders and are ethically sound results in ethical governance.

Acceptance of worldwide indicators for ethical governance

The worldwide indicators given by the WB which must be accepted for ethical governance are, namely, the following:

- Voice and accountability
- Political stability and absence of violence
- Government effectiveness
- Regulatory quality
- Rule of law
- Control of corruption[1]

Since most developing countries today are ridden with corruption, they are still struggling with development issues. Countries that rank low on the corruption index tend to deteriorate their legitimacy from a global perspective. Good governance based on ethics liberates the society to live to its best capacity. The World Bank indicators included ethical governance characteristics such as the kind of political regime, methodology of economic and social resources management and capabilities of governments in policy formulation and execution.

The most crucial factor impeding good governance is corruption. The factors leading to corruption are being studied globally by economists and thinkers. Corruption slows growth, whether it is economic or social. The state has to

---

1 World Bank, *Worldwide Governance Indicators*, <https://www.worldbank.org/en/publication/worldwide-governance-indicators> (accessed 10 August 2024).

intervene to sever the corrupt relationship between the governed and the governing authority and build better systems that will not let corruption permeate the organization. Civil society's role has to be considered in order to achieve the goal of good governance. Citizen participation without any political alignment is necessary. Anticorruption strategies and policies are the tools to combat the lack of ethics. Good governance needs to investigate the issue of work ethics. The office of the bureaucracy lacks work responsibility, work conscience and ethical work conduct. Commitment to work cannot be bought but needs to be inculcated as it is elusive, and there is no method to achieve it.

Good governance and corruption

Meaning and concept of corruption

Though various organizations have given many definitions worldwide, the issue here is understanding corruption rather than defining it. Both concepts must be understood to understand the link between good governance and corruption. Corruption is an improper use of mandated authority by the elected representatives or the executive. The World Bank (WB) defines corruption 'as abuse of public power for private benefit'. Corruption can occur between individuals and business concerns. It commonly occurs when large amounts of money are involved or when aid is partaken by people experiencing poverty and developing nations from the developed ones, in an e-commerce environment, natural or man-made disasters and many other similar situations. It is found everywhere in most countries and is a part of daily life. It becomes massive with more significant amounts of money and higher positions of power. Bribery, extortion, nepotism, fraudulent embezzlement and kickbacks to political parties are some of its aspects.

Good governance happens where there is no corruption, and development only occurs when anticorruption moves on to become an intrinsic part of good governance. The IMF strongly emphasises good governance by minimizing and eliminating corruption. As corruption leads to mistrust in government among the public, it becomes detrimental to progress and growth in the economy and market. The IMF considered this so important that in 1997, it developed a policy named 'The Role of the IMF in Governance Issues' to tackle economic governance. The IMF has maintained a relationship with its member nations to fight corruption. It promotes surveillance in financial and technical help so that a secure environment limits corruption. It also checks money laundering. ADB defines corruption as abusing public or private office for personal gain. This points to any behaviour in which people in the public or private sector improperly and unlawfully enrich themselves or those close to them or induce others to do so by misusing their position.

The Governance and Anticorruption (GAC) Strategy

The World Bank reaffirmed its commitment in September 2007 to confront corruption as a core development issue wherever it exists and support integrity in public sector institutions. It agreed to the following:

- Develop the capacity of the country's clients to deliver on their commitments to enhance transparency and reduce corruption
- Enhance its support for the implementation of anti-money laundering requirements and the recovery of stolen assets
- Extend its work on tax reform, IFFs, procurement reform and preventing corrupt companies from winning state contracts.

The World Bank (WB) has played a key role in developing and executing the Governance and Anticorruption (GAC) Strategy. Its primary goal is to improve its documentation system and manage loans more efficiently. This approach would enhance the effectiveness of the WB's system if its organizational structure is tailored to meet its specific needs. The GAC Strategy tabled a report based on its achievements. The GAC Strategy has some objectives on its agenda, which can be summarised as follows:

1. Reduction in poverty: Poverty reduction is one of GAC Strategy's primary objectives by having an environment of better services and infrastructure catering to the following four aspects:
   - Country assistance strategies: The GAC Strategy focused on developing tailored strategies to meet each country's unique needs, ensuring the creation of effective and sustainable development infrastructure.
   - Sectoral strategies: Within the scope of the GAC Strategy, it finds plausible methods to promote a particular sector.
   - Project design and implementation: This formulates sectoral strategies to get better transparency, citizen participation, supervision and judicial monitoring by some neutral body and some control of the contract of the project functioning design.
   - Country systems: The GAC Strategy aims at developing country-specific development of individuals' capacities in all projects.
2. Twofold manner: The GAC Strategy works in a twofold manner. On the one hand, it strives to revamp the methodology of coping with developmental governance issues concerning effectiveness. On the other hand, it moves toward tackling the GAC Strategy issues.
3. Designing programmes: The GAC Strategy designs and executes developmental programmes in a way that they deal with political dilemmas

efficiently. So, it aims to have a sound system in sync with its strategies, dwelling seriously on the political decision-making of that particular country and promoting capacity strength and monitoring.

4. Risk management: The GAC Strategy helps manage risk as it became clear after the 2008 economic crisis. It learnt how such factors need to be included in the GAC Strategy to do better work.
5. Country-specific issues: The GAC Strategy must deal with country-specific problems regarding corruption in the path of developmental effectiveness. Risk management, and not avoiding risks, is the solution. The poorest nations need confidence-building at the grass-roots level by establishing an institutional framework and infrastructure. Building nations is not risk-free, and the GAC Strategy must ensure the World Bank's high standards are upheld successfully.
6. Evaluation of GAC Strategy implementation: GAC Strategy must assess its progress in executing its programmes and projects. It assesses two ways:
   - Assessment of the limits of achievement of GAC Strategy in numbers that have improved World Bank development tasks.
   - Assessment of the effect of GAC Strategy work on the effectiveness of development projects.
7. Link between the World Bank's aims and the GAC Strategy: The GAC Strategy needs to be implemented to promote the World Bank's aims. The country-specific strategies must be in the World Bank's aims to make development visible at the grass-roots level. It stresses 'change management' efforts to execute the World Bank Group's (WBG) strategies for good project designs and implementation while checking corruption. Core country systems for procurement, management and decentralization are being strengthened. It stresses the 'who' and 'how' of having sufficient execution of the GAC Strategy skills and information.

Interlinkage between corruption and growth

Social scientists are doing research to understand the effects of corruption. On the one hand, one set of researchers says that the more corrupt a country is, the lower the curve of its economic growth. On the other hand, after studying Indonesia's economy, some researchers State that institutionalised corruption is better than non-institutional corruption. They also say it may improve efficiency and growth as it does away with the government-imposed rules and limits. They also promote the idea that the most efficient firms pay the highest bribes, which ensures that the best company gets the project. Decisions can be made quicker, and bribes can keep the officials happy. Strangely enough, research that propagates corruption cannot be supported.

First, the arguments for corruption can be refuted because rules and limits are primarily made to make it easier to get bribes. Second, it's possible that the highest bribe payers aren't the most efficient, and officials might deliberately slow down work unless bribes are paid. On a qualitative level, corruption decreases public revenue. It leads to an increase in public spending, which is adverse to the deficits, and the government faces a problem in having a stable fiscal policy. Economic inequality becomes common because of bribes. Market economies move the way the corruption wants them to, straying away from the course of the government's plan.

After studying various aspects of good governance, one can be safely say that a nation cannot be run without good governance. Presently, a lot of stress is being laid on good governance to run the significant sectors of the social infrastructure, such as health and education.

Sustainable development and corruption: World Bank's perspective

The World Bank Group considers corruption a major challenge to its twin goals of ending poverty by 2030 and boosting shared prosperity for 40 per cent of the population in developing countries. The focus of the SDGs is to reduce poverty and to achieve the targets set for financing development.[2] According to the World Bank, businesses and individuals pay 1.5 trillion dollars in bribes annually, about 2 per cent of the global GDP and ten times the amount of overseas development assistance. Corruption causes a negative impact, which is multifold on people with low incomes and on economic growth due to its enormous volume. Sustainable Development takes a brutal hit due to money being laundered as bribes and ending in the wrong pockets.

The following are the negative impacts that corruption has on the achievement of SDGs:

- Health: Poor people are deprived of access to health services, and this has adverse effects on the infant mortality rate.
- Crimes: When large sums of money are lost through IFFs, this money finds a way to fund drugs and human trafficking. Corruption gives rise to crimes of the highest heinous extent, as well as terrorism. If governance is improved by inculcating ethical surpluses, there can be a 300 per cent increase in dividends from making weak governance stronger.

2 World Bank, *Anticorruption for Development*, <https://www.worldbank.org/en/programs/anticorruption-for-development>, accessed 10 August 2024.

- Economic activity: Corruption acts as a powerful disincentive to foreign investment. Countries with weak ethical governance have been empirically found to be underdeveloped or developing, and those that have managed to curb corruption have attracted more foreign and domestic investment.
- Service delivery: Good governance rests upon an efficient public service delivery system, but the ethical deficit that leads to corruption hampers this process. People experiencing poverty are left poorer and the rates of mortality and diseases go higher. Educational standards are found to be inadequate in quality. The police take bribes to manipulate their power and authority by putting funds into their pockets and helping government officials swindle accounts.
- Institutional hazards: Corruption pervades the institutional framework, and its mechanisms fail to deliver honesty. Projects remain unfinished, causing a setback in development. The institutions are corrupt in the underdeveloped Southern Hemisphere, and money is being transferred from poor to wealthy countries to undermine development.

Reasons for corruption in governance

Corruption is the main reason why ethical governance faces issues in its implementation. There is no shortage of funding for development projects and schemes, especially in India. One look at the union budgets of post-independence India shows that the budget allocations are only for the development work for the socio-economic welfare of the population of India. However, despite seven decades of efforts, India ranks low in all the development indices. India is a rapidly emerging market economy, and one can easily see the right direction it has taken. Still, low-income people remain poorer, resulting in a more significant socio-economic disparity. Two reasons that come to the forefront after a careful analysis of the economic surveys of the Indian budget and budgetary appropriation of funds are: first, lack of objective and the will to implement schemes and second, corruption.

It is understood that corruption is the major deterrent in ethical governance to develop our nation, so the reasons for corruption must be understood. Why do people indulge in corrupt practices? Why do officials and businesspersons indulge in bribery-related illegal transactions? Gary Becker, the Nobel Laureate in Economics, opined that corruption can be eradicated solely by abolishing the State. Since this cannot happen, solutions to this problem need to be studied. Countries like Canada, Finland, Denmark, the Netherlands and Sweden have vast PSUs but the least corruption measures, so how the state functions is more important than the size of the government.

Sustainable development and ethics

Sustainable Development Goal (SDG) 16: Peace, Justice and Strong Institutions is the SDG most closely related to ethical sustainable governance. SDG 16 aims to 'promote peaceful and inclusive societies for sustainable development, provide access to justice for all, and build effective, accountable, and inclusive institutions at all levels'.

Key targets under SDG 16 that relate to ethical governance include:

- 16.3 Promote the rule of law at the national and international levels and ensure equal access to justice for all.
- 16.5 Substantially reduce corruption and bribery in all their forms.
- 16.6 Develop effective, accountable, and transparent institutions at all levels.
- 16.7 Ensure responsive, inclusive, participatory, and representative decision-making at all levels.
- 16.10 Ensure public access to information and protect fundamental freedoms.

These targets emphasise the importance of governance structures that are ethical, transparent, accountable and inclusive, which are all essential components of sustainable development governance. SDG 16 recognises that strong institutions and ethical governance are critical to achieving sustainable development and ensuring its benefits are distributed equitably and justly across all segments of society.[3]

The attainment of sustainable development governance, good governance and ethics globally involves a multifaceted approach that includes international cooperation, establishing global standards, and implementing robust governance frameworks. This effort is being pursued through various global initiatives, agreements, and institutions, which increasingly recognise the importance of integrating ethical principles into governance and sustainable development practices.[4]

1. International cooperation and agreements: Globally, the United Nations' 2030 Agenda for Sustainable Development, which outlines the Sustainable

3 United Nations, *Transforming Our World: The 2030 Agenda for Sustainable Development* (New York: United Nations, 2015), <https://sdgs.un.org/goals/goal16>, accessed 10 August 2024.

4 United Nations Development Programme, *Goal 16: Peace, Justice, and Strong Institutions* (New York: United Nations Development Programme, 2021), <https://www.undp.org/sustainable-development-goals/goal-16-peace-justice-and-strong-institutions>, accessed 10 August 2024.

Development Goals (SDGs), is a cornerstone for aligning national and international policies with sustainability objectives. The SDGs emphasise the need for inclusive, transparent, and accountable governance structures, highlighting ethical considerations such as equity, justice, and environmental stewardship. International agreements like the Paris Agreement on climate change further reflect global commitments to sustainable development, driven by ethical imperatives to protect the planet for future generations.

2. Global standards and institutions: Institutions such as the United Nations, the World Bank and the International Monetary Fund (IMF) are critical in promoting sustainable development governance by setting global standards and guiding member states. These institutions advocate for governance practices that align with ethical principles, including transparency, accountability and participation, which are essential for sustainable development. Organizations like the Organisation for Economic Cooperation and Development (OECD) also work with countries to implement governance reforms that promote sustainable economic growth while ensuring social and environmental well-being.
3. National implementation and policy integration: Countries worldwide increasingly integrate sustainable development governance into their national policies. This involves adopting ethical governance practices that promote transparency, inclusivity, and accountability in decision-making processes. For example, many countries have established national strategies for implementing the SDGs, including frameworks for monitoring and reporting progress, ensuring that policies align with sustainability goals and ethical standards.
4. Private sector and civil society engagement: The private sector and civil society organizations (CSOs) are also crucial in attaining global sustainable development governance. Businesses are increasingly adopting corporate social responsibility (CSR) practices and sustainability reporting, driven by ethical considerations and consumer demand for more responsible business practices. CSOs are vital in holding governments and corporations accountable, advocating for ethical governance practices and prioritizing social and environmental sustainability.
5. Ethical leadership and education: Promoting ethical leadership and education is essential for achieving sustainable development governance globally. Leaders at all levels are encouraged to prioritise ethical considerations in their decision-making processes, ensuring their actions contribute to long-term sustainability. Educational initiatives, both formal and informal, are crucial for raising awareness about the importance of ethical governance

and sustainability, empowering individuals and communities to advocate for and participate in governance processes.

Thus, the global attainment of sustainable development governance, good governance, and ethics is a complex and ongoing process that requires collaboration among international organizations, governments, the private sector, and civil society. By aligning governance practices with ethical principles, the global community is working towards a more sustainable and equitable future for all.

Ethical sustainable development governance models

Several specific models and practices on ethics are widely recognised and implemented globally across various sectors. These models serve as frameworks to guide ethical decision-making and behaviour in governance, business and other areas. Here are some notable examples:

1. **The Ethical Decision-Making Model:** This model is often used in organizations to help individuals and groups make ethical decisions. The Ethical Decision-Making Model is a structured approach used to guide individuals and organizations through making ethical decisions that align with their values. The model typically consists of several key steps, each encouraging careful consideration of the ethical implications of potential decisions.[5]

Steps in The Ethical Decision-Making Model

Given below is the process involved in the Ethical Decision-Making Model in Figure 33 and described briefly:

- Identify the ethical issue: The first step involves recognizing that a situation has ethical implications. This might include conflicts of values, potential harm to stakeholders or decisions that could challenge an individual's or organization's integrity.
- Gather relevant information: Gathering all relevant facts before deciding is essential. This includes understanding the context, identifying the stakeholders involved and considering all applicable laws, policies, or guidelines.
- Evaluate the alternatives: All available options available for action must be considered. Each alternative should be evaluated in light of ethical principles, such as honesty, fairness, respect and responsibility. Ethical theories, such as

5 O. C. Ferrell, John Fraedrich, and Linda Ferrell, *Business Ethics: Ethical Decision Making and Cases* (Boston, MA: Cengage Learning, 2015).

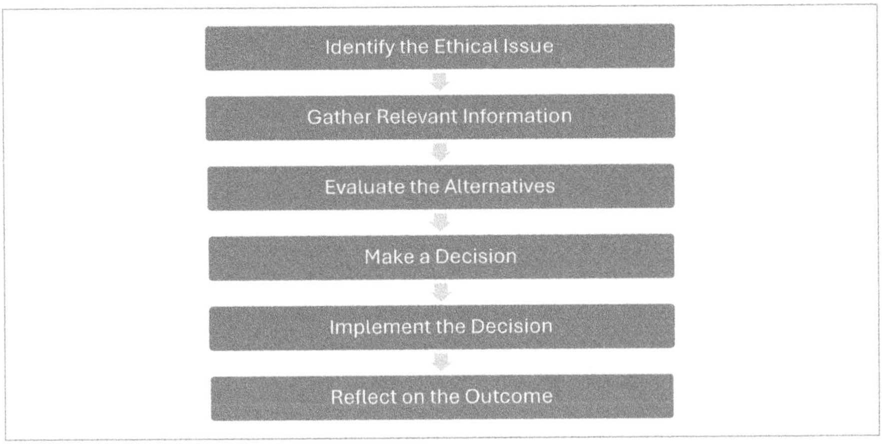

Figure 33 The ethical decision-making model

utilitarianism (which focuses on outcomes) and deontology (which focuses on duties and rules), can be applied to weigh the pros and cons of each option.

- Make a decision: After evaluating the alternatives, the decision-maker must choose the course of action that best aligns with ethical principles and the organization's values. This step often involves balancing competing interests and making trade-offs.
- Implement the decision: Once a decision has been made, it must be implemented. This involves planning the decision's execution, communicating with stakeholders and ensuring the necessary resources and support are available.
- Reflect on the outcome: After the decision has been implemented, it's essential to reflect on the outcome using any monitoring and evaluation best practices. This reflection should consider whether the decision achieved the desired ethical outcome, what went well, and what could be improved in future decisions.[6]

Importance: The Ethical Decision-Making Model is crucial because it provides a clear and systematic way to address ethical dilemmas. It encourages individuals and organizations to consider the broader impact of their decisions, fosters transparency and helps ensure that actions are consistent with ethical standards. By following this model, decision-makers can avoid ethical pitfalls, build trust with stakeholders, and contribute to a culture of integrity and accountability.

6 Linda K. Treviño and Katherine A. Nelson, *Managing Business Ethics: Straight Talk about How to Do It Right* (Hoboken, NJ: Wiley, 2016).

2. **The Triple Bottom Line (TBL) Model:** This sustainability framework evaluates a company's performance based on three criteria: people (social impact), planet (environmental impact) and profit (economic impact). The TBL approach encourages businesses to go beyond the traditional financial metrics and consider the broader impact of their operations on society and the environment. This aligns with ethical business practices and promotes corporate responsibility. This model encourages businesses and organizations to focus on profit and social and environmental impacts. The model was introduced by John Elkington in 1994 and has since become a widely adopted concept in sustainability practices.[7]

Critical Components of the TBL Model

The Triple Bottom Line (TBL) model, which integrates social, environmental, and economic considerations in sustainable development, relies on several critical components that ensure a holistic approach to organizational success. They are:

- People (Social Bottom Line): The component of 'people' emphasises an organization's impact on all the people involved or affected by its operations. It includes employees, customers, suppliers and the community at large. Companies are encouraged to operate in ways that benefit society by promoting fair labour practices, ensuring health and safety, contributing to social well-being and engaging in community development activities.
- Planet (Environmental Bottom Line): The environmental bottom line focuses on an organization's ecological footprint. This involves minimizing the negative environmental impacts of business activities, such as reducing waste, conserving energy, and using resources sustainably. The goal is to ensure that business operations do not deplete natural resources or harm the environment and to promote practices that contribute to environmental sustainability.
- Profit (Economic Bottom Line): While traditional business models focus solely on profit, the TBL model redefines profit in a broader context. The economic bottom line includes not just financial performance but also the organisation's economic impact on the community and society as a whole. This includes creating jobs, stimulating economic growth, and ensuring long-term financial stability while considering economic activities' social and environmental consequences.

---

7 John Elkington, *Cannibals with Forks: The Triple Bottom Line of 21st-Century Business* (Gabriola Island, BC: New Society Publishers, 1998).

Importance: The TBL model provides a comprehensive approach to sustainability by focusing on the three pillars: people, planet and profit. Organizations can make more balanced decisions and align with long-term sustainability goals. The model encourages businesses to consider the broader impact of their operations, leading to more responsible and ethical decision-making. While the TBL model is widely praised, it has also faced criticism. Some argue that measuring social and environmental impacts as accurately as financial performance is complex. Balancing the three components can be challenging, especially when economic goals conflict with social or environmental objectives. This has always been the case in development administration, where the trade-off between development and sustainability was unequal. Despite these challenges, the TBL model remains a crucial framework for businesses aiming to achieve sustainable development.[8]

3. **Stakeholder Theory:** This theory, developed by R. Edward Freeman in his 1984 book *Strategic Management: A Stakeholder Approach*, posits that businesses and organizations have a responsibility to shareholders and all stakeholders affected by the organization's activities. This includes employees, customers, suppliers, communities and the environment. Stakeholder theory advocates for ethical decision-making that considers the interests of all these groups, promoting fairness and responsibility. The theory argues that a company's success depends on its ability to create value for all its stakeholders, not just shareholders.[9]

Critical Elements of Stakeholder Theory: This theory has five primary elements as the foundation, which are as follows.

- Definition of stakeholders: According to Freeman, stakeholders are 'any group or individual who can affect or is affected by the achievement of an organization's objectives'. This broad definition includes shareholders, employees, customers, suppliers, communities, governments and even competitors.
- Stakeholder interests: Companies should consider the interests and rights of all stakeholders in their decision-making processes. This involves balancing the sometimes competing needs of different stakeholders, such as profit for shareholders, fair wages for employees, product safety for customers, and environmental protection for the community.

8 Wayne Norman and Chris MacDonald, 'Getting to the bottom of the "Triple Bottom Line"', *Business Ethics Quarterly* 14, no. 2 (2004): 243–262.

9 R. Edward Freeman, *Strategic Management: A Stakeholder Approach* (Boston, MA: Pitman, 1984).

- Ethical and moral foundations: The theory is grounded in ethics and suggests that businesses have a moral obligation to consider the impact of their actions on all stakeholders. It challenges the traditional view that a company's primary responsibility is to maximise shareholder value, instead advocating for a more inclusive approach to business ethics.
- Stakeholder engagement: Engaging with stakeholders is a crucial component of Stakeholder Theory. This implies open communication, transparency, and building relationships with stakeholders to understand their concerns and expectations. Companies that successfully engage with stakeholders can build trust, enhance their reputation, and create long-term value.
- Value creation: Businesses create value by serving the interests of all stakeholders, not just shareholders. When a company considers the needs of its stakeholders, it can make a more sustainable and resilient business model. This holistic approach often leads to better decision-making and long-term success.[10]
- Applications and implications: Stakeholder Theory influences corporate governance practices by encouraging companies to include stakeholder interests in their governance structures. This might involve having stakeholder representatives on boards or adopting policies that promote stakeholder engagement. The theory is closely related to concepts like Corporate Social Responsibility (CSR) and sustainability, as it encourages companies to consider their operations' social and environmental impacts. From a strategic perspective, companies that adopt Stakeholder Theory often focus on long-term value creation rather than short-term profits. This can lead to more sustainable business practices and improved relationships with stakeholders.[11]

While Stakeholder Theory is widely accepted and has influenced modern business practices, it is not without criticism. Some argue that it is challenging to balance the interests of all stakeholders, particularly when those interests conflict. Others contend that the theory lacks clarity in defining who qualifies as a stakeholder and how much weight their interests should carry in decision-making.

4. The Code of ethics: Many organizations and professions have established codes of ethics formal documents outlining the principles and standards of ethical behaviour expected from members. Examples include the

10 Thomas Donaldson and Lee E. Preston, 'The stakeholder theory of the corporation: Concepts, evidence, and implications', *Academy of Management Review* 20.1 (1995): 65–91.

11 Robert Phillips, *Stakeholder Theory and Organizational Ethics* (San Francisco, CA: Berrett-Koehler Publishers, 2003).

International Federation of Accountants' Code of Ethics, The American Medical Association's Code of Medical Ethics and The United Nations Global Compact for Businesses.[12] These codes typically cover integrity, accountability, transparency, and respect for human rights, providing a clear ethical framework for professionals and organizations. A Code of Ethics is a set of principles and guidelines designed to help professionals conduct their business with honesty, integrity, and fairness. It serves as a framework for ethical decision-making and behaviour, outlining the standards individuals or organizations must adhere to in their professional activities. Codes of ethics are commonly adopted by organizations, professional associations, and institutions to ensure that their members act per the values and principles that define the organization.

Components of a code of ethics

- Core values: A code of ethics typically begins by outlining the organisation's or profession's core values. These values might include honesty, integrity, accountability, respect, fairness and responsibility. These values serve as the foundation upon which the specific ethical guidelines are built.
- Principles and standards: The code will specify the ethical principles that guide the behaviour of individuals within the organization. This can include principles related to confidentiality, transparency, conflicts of interest, professional competence and respect for others. The code may also include specific standards of conduct that professionals must follow, such as avoiding fraudulent activities or maintaining impartiality in decision-making.
- Compliance and enforcement: A code of ethics often includes provisions for enforcing its ethical guidelines. This can involve mechanisms for reporting unethical behaviour, procedures for investigating complaints, and consequences for violating the code. Some organizations may have ethics committees or officers responsible for ensuring compliance with the code.
- Commitment to stakeholders: Many codes of ethics emphasise the responsibility that professionals have towards various stakeholders, including clients, customers, colleagues, and the wider community. This aspect of the code ensures that the actions of the professionals contribute positively to society and do not harm others.

12 American Medical Association, *AMA Code of Medical Ethics*, <https://www.ama-assn.org/delivering-care/ethics/code-medical-ethics-overview>, accessed 10 August 2024.

Importance of a code of ethics

A code of ethics may provide a decision-making framework to help individuals resolve ethical dilemmas. This might include identifying the ethical issue, considering the possible consequences of different actions, and consulting with colleagues or ethics committees before deciding. A few key points of importance of a code of ethics are:

- Guidance: A code of ethics provides clear guidance on the expected conduct of professionals, helping them navigate complex situations and make decisions that align with the organization's values.
- Accountability: It holds individuals accountable for their actions by setting standards that must be met, thus maintaining the integrity and reputation of the profession or organization.
- Trust: A well-enforced code of ethics fosters trust among stakeholders, including clients, employees, and the public, by demonstrating the organization's commitment to ethical behaviour.
- Consistency: It ensures consistency in behaviour and decision-making across the organization, reducing the risk of unethical practices and promoting a positive workplace culture.
- Some examples of Codes of Ethics are The American Medical Association (AMA) Code of Medical Ethics and The International Federation of Accountants (IFAC) Code of Ethics for Professional Accountants.

5. Corporate Social Responsibility (CSR) Practices: CSR involves companies taking responsibility for their impact on society and the environment. Ethical practices under CSR include philanthropy, ethical labour practices, environmental sustainability efforts and community engagement. Many companies implement CSR strategies to demonstrate their commitment to ethical principles and contribute positively to society. CSR refers to the practices and policies businesses adopt to manage their impact on society and the environment. It encompasses various activities, from philanthropic efforts to ethical labour practices, environmental sustainability initiatives, and economic contributions. The concept is rooted in the belief that companies have a responsibility not only to their shareholders but also to other stakeholders, including employees, customers, communities and the environment.[13]

---

13 Archie B. Carroll, 'The pyramid of corporate social responsibility: Toward the moral management of organizational stakeholders', *Business Horizons* 34.4 (1991): 39–48.

Principal areas of CSR practices

The principal areas of Corporate Social Responsibility (CSR) practices encompass a wide range of initiatives focused on ethical governance, environmental sustainability and social equity, which collectively drive responsible business operations and community engagement. They are:

- Environmental responsibility: Many companies engage in sustainability practices such as reducing carbon footprints, using renewable energy, and promoting waste reduction and recycling. These efforts are aimed at minimizing environmental impact and promoting ecological balance. Companies may develop and promote environmentally friendly products or services, called green products. This includes using sustainable materials, reducing packaging and ensuring products are energy-efficient or biodegradable.
- Social responsibility: Businesses often support local communities through charitable donations, volunteerism, and partnerships with non-profit organizations to encourage community engagement. This might involve sponsoring local events, funding community projects, or providing resources to those in need. Secondly, CSR includes ensuring fair labour practices, promoting diversity and inclusion, providing a safe working environment, and offering opportunities for employee development and well-being.
- Economic responsibility: Companies engaged in CSR often focus on conducting business ethically, including fair trade practices, transparency in financial reporting and an environment with anticorruption measures. Ensuring the supply chain is ethical, including sourcing materials from suppliers who adhere to fair labour practices and environmental standards.
- Corporate governance: CSR involves maintaining high standards of corporate governance, which includes transparency in decision-making, accountability to stakeholders, and ethical conduct at all levels of the organization. Companies practising CSR actively engage with their stakeholders, including customers, employees, investors, suppliers and the community, to understand their concerns and incorporate feedback into business practices.

Examples of CSR practices:

- Environmental stewardship: Patagonia, an outdoor clothing company, is known for its commitment to environmental sustainability. The company donates a portion of its profits to environmental causes, uses recycled materials in its products and encourages customers to repair and recycle their gear.
- Social programs: Starbucks has implemented numerous CSR initiatives focused on social responsibility. These include providing healthcare benefits to employees and promoting ethical sourcing of coffee beans through its

Coffee and Farmer Equity (CAFE.). It practices this program and supports local communities through charitable contributions.

- Philanthropy: Microsoft engages in CSR through its philanthropic efforts, including the Microsoft Philanthropies program, which donates software, services and cash to non-profits and educational institutions worldwide. The company also encourages employee volunteerism and matches charitable contributions made by its employees.

Importance of CSR: CSR can enhance a company's reputation and build trust with consumers, employees, and other stakeholders. Companies perceived as socially responsible may attract customers and retain employees more effectively. Secondly, CSR can help companies mitigate risks, such as regulatory compliance issues or negative publicity, by proactively addressing social and environmental concerns. Lastly, by focusing on sustainable practices, companies can ensure their long-term viability and contribute to society's overall well-being.

Challenges in implementing CSR

- Balancing profit and responsibility: Companies often face challenges in balancing their profit motives with their CSR commitments, particularly when these objectives conflict.
- Measuring impact: It can be difficult to quantify the impact of CSR initiatives, making it challenging to assess their effectiveness and return on investment.
- Stakeholder expectations: Managing diverse stakeholder expectations can be complex, as different groups may have varying priorities and views on what constitutes responsible corporate behaviour.

5. The Global Reporting Initiative (GRI): The GRI is an international standards organization that helps businesses, governments, and other organizations understand and communicate their impact on critical sustainability issues. The GRI Standards offer a universal framework for reporting economic, environmental and social effects, promoting transparency and ethical behaviour in organizations worldwide. It is an international independent standards organization that helps businesses, governments, and other organizations understand and communicate their impacts on climate change, human rights, and corruption. GRI provides a comprehensive sustainability reporting framework widely used worldwide to guide

organizations in disclosing their environmental, social, and governance (ESG) performance.[14]

Critical aspects of the GRI

The critical aspects of GRI are given below:

- GRI Standards: The GRI Standards are the most widely adopted global standards for sustainability reporting. They offer a modular, interrelated structure that allows organizations to report on various topics, including economic performance, environmental impact, labour practices, human rights and societal impact. The standards are designed to be used by organizations of any size, sector or location.
- Materiality: One of the core principles of the GRI Standards is materiality. This concept focuses on the importance of reporting information that reflects the organization's significant economic, environmental, and social impacts or that could influence the decisions of stakeholders. Organizations are encouraged to engage with stakeholders to identify what issues are material to them and report on those aspects.
- Transparency and accountability: The GRI encourages organizations to disclose both positive and negative aspects of their performance. This balanced reporting allows stakeholders to assess an organization's sustainability efforts. Accountability is emphasised through the requirement for organizations to explain how they manage their impacts and their actions to improve their performance.
- Stakeholder inclusiveness: The GRI Standards advocate for stakeholder inclusiveness, meaning that organizations should identify their stakeholders, including employees, customers, suppliers, communities and investors. Stakeholders should explain how they have responded to their reasonable expectations and interests. This helps ensure that the sustainability report addresses the concerns of those affected by the organization's activities.
- Global adoption and impact: The GRI Standards are widely used across various sectors globally, making them a critical tool for benchmarking and improving sustainability practices. Organizations adopting GRI Standards can contribute to global sustainability goals, such as the United Nations Sustainable Development Goals (SDGs).

14 Global Reporting Initiative, *GRI Standards* (2023), <https://www.globalreporting.org/standards>, accessed 10 August 2024.

The importance of GRI Standards is in how they provide a consistent format for reporting, making it easier for stakeholders to compare the sustainability performance of different organizations. This comparability enhances accountability and drives improvements across industries. Secondly, reporting according to GRI Standards can strengthen the credibility of an organization's sustainability efforts, as the standards are recognised globally and are developed through a rigorous multi-stakeholder process. Finally, organizations that align their strategies with sustainability reporting gain a strategic advantage by anticipating and managing risks, responding to stakeholder demands and identifying opportunities for innovation. Some examples of organizations that use GRI Standards are Unilever, Volkswagen and Coca-Cola. But GRI is not without its challenges. Its breadth and depth can make reporting complex and resource-intensive, especially for smaller organizations. As sustainability issues evolve, the GRI Standards are periodically updated, requiring organizations to stay current with the latest guidelines and adapt their reporting practices accordingly.

6. Ethical leadership models: Ethical leadership is a model where leaders prioritise ethical standards in their decision-making and behaviour, influencing their organization's culture. This involves leading by example, promoting an ethical culture and ensuring organizational practices align with ethical values. Ethical leaders are seen as role models who foster trust, fairness and accountability within their organizations. Ethical leadership is crucial because it sets the tone for the behaviour of the entire organization and influences how employees perceive and act on ethical issues.

Key Ethical Leadership Models

Key ethical leadership models provide frameworks for understanding how leaders can navigate complex moral landscapes, inspire ethical behaviour in their organizations, and contribute to a culture of integrity and social responsibility. They are:

- Transformational leadership: Transformational leaders inspire and motivate followers to achieve higher performance while focusing on ethical standards. They emphasise the importance of vision, integrity, and personal development, often encouraging followers to transcend their self-interests for the organisation's or society's good. Transformational leaders are usually seen as role models who embody high ethical standards, influencing their followers to adopt similar values. Their commitment to ethics is reflected in their decision-making processes and interactions with others. For example,

Nelson Mandela is often cited as an example of a transformational leader who emphasised reconciliation, integrity, and social justice.

- Servant leadership: Servant leadership prioritises the needs of others, particularly employees and other stakeholders, before the leader's interests. This model is based on the belief that the primary purpose of leadership is to serve. The ethical focus is that the Servant leaders are committed to the growth and well-being of individuals and communities. They foster an environment where ethical behaviour is encouraged and supported and focus on building trust, empathy, and community within the organization. For example, Leaders like Mahatma Gandhi and Mother Teresa exemplify servant leadership, focusing on serving others and promoting social equity.
- Authentic leadership: Authentic leadership is centred around the leader's self-awareness, relational transparency and consistency in upholding ethical standards. Authentic leaders are genuine, self-regulated and lead by example. Authentic leaders build credibility and trust by following their values and ethical principles. They are transparent in their actions and decisions, which fosters an open and ethical organizational culture. For example, Howard Schultz, the former CEO of Starbucks, has been described as an authentic leader due to his focus on transparency, ethical sourcing, and social responsibility.
- Moral leadership: Moral leadership is based on the principle that leaders should act according to universal ethical standards, such as justice, fairness and respect for human dignity. Moral leaders prioritise ethical considerations over personal or organizational gain. Moral leaders emphasise the importance of ethical reasoning and decision-making. They are committed to doing what is right, even when difficult or unpopular. For example, Martin Luther King Jr. is an iconic moral leader who advocated for civil rights based on the principles of justice and equality.
- Ethical decision-making models: Ethical decision-making models provide a structured approach for leaders to assess the ethical implications of their decisions. These models typically involve identifying the ethical issue, considering the stakeholders, evaluating alternatives and making decisions that align with ethical principles. These models help leaders navigate complex ethical dilemmas by providing a transparent process for weighing the consequences of different actions and choosing the most ethical course. For example, The PLUS model, which stands for Policies, Legal, Universal, and Self, is one ethical decision-making framework that helps leaders evaluate the ethical aspects of their decisions.

Importance of ethical leadership models: Ethical leadership is essential for building trust within and with external stakeholders. Trustworthy leaders create an environment where employees feel safe and are more likely to engage in ethical behaviour. Leaders can influence the organizational culture, encouraging employees to act ethically and report unethical practices by modelling ethical behaviour. Organizations led by ethical leaders are more likely to achieve long-term success because they prioritise sustainable practices, stakeholder relationships, and social responsibility.

However, these models do not always run smoothly, as ethical leaders often face the challenge of balancing the interests of various stakeholders, which may sometimes be in conflict. Leaders may find it challenging to maintain their ethical standards in high-pressure environments, especially when faced with short-term goals or external pressures.

7. ISO 26000 - Social responsibility guidance is an international standard developed by the International Organization for Standardization (ISO) to provide guidelines for organizations on social responsibility. Unlike other ISO standards, ISO 26000 is not certifiable, meaning organizations cannot be certified for compliance. Instead, it offers guidance on how businesses and organizations can operate in a socially responsible way, contributing to sustainable development.

Fundamental Aspects of ISO 26000

The key aspects of ISO 26000 are given below:

1. Core Subjects: ISO 26000 identifies seven core subjects of social responsibility:
   - Organizational governance: Effective governance structures to integrate social responsibility.
   - Human rights: Respect and protect human rights in all organizational activities.
   - Labour practices: Promote fair labour practices, including workplace health and safety.
   - Environment: Minimise environmental impact through sustainable practices.
   - Fair operating practices: Encourage ethical conduct, anticorruption and responsible political involvement.
   - Consumer issues: Promote fair marketing, health and safety and sustainable consumption.
   - Community involvement and development: Engage in community activities that foster development and social well-being.

2. Principles of social responsibility: The standard outlines several principles that should guide organizations, including accountability, transparency, ethical behaviour, respect for stakeholder interests, respect for the rule of law, respect for international norms of behaviour and respect for human rights.
3. Stakeholder engagement: ISO 26000 emphasises identifying and engaging with stakeholders to understand their expectations and address their concerns effectively.
4. Integration into strategy: The guidance stresses that social responsibility should be integrated into an organization's strategy, operations and decision-making processes to implement it effectively.
5. Global relevance: ISO 26000 is designed to apply to all types of organizations, regardless of size, location, or industry. It considers diverse cultural, social, environmental, and legal contexts.

Benefits of implementing ISO 26000

Adopting ISO 26000 can enhance an organization's reputation by demonstrating a commitment to ethical behaviour and social responsibility. Organizations can reduce risks associated with unethical practices, legal liabilities and damage to reputation by addressing social responsibility issues. Engaging stakeholders and addressing their concerns can lead to stronger relationships and greater trust. Finally, Implementing ISO 26000 helps organizations contribute to achieving the United Nations Sustainable Development Goals (SDGs), particularly those related to social and environmental sustainability. Nonetheless, it is not without challenges due to the complexity of its implementation. As a non-certifiable standard, it is voluntary, and there is no external compliance verification, which may limit its adoption and impact. It defines social responsibility and helps organizations translate principles into practical actions, promoting ethical practices in human rights, labour practices, the environment, fair operating practices and community involvement. These models and practices are implemented globally to promote ethical behaviour and decision-making across various sectors, contributing to sustainable development and good governance by ensuring ethical considerations are at the forefront of organizational and governmental actions.

Conclusion

A code of conduct is a set of rules that outlines the responsibilities of individuals or organizations, guiding behaviour based on ethical principles. These codes are essential for public servants, providing clear guidelines for actions and ensuring accountability and efficiency. The International Federation of Accountants defines a code of conduct as principles and standards contributing to stakeholder welfare and respecting the rights of all those affected by an organization's operations. The importance of ethical codes lies in their role in guiding government officials and public administrators. Without such guidelines, corruption can undermine developmental efforts. Ethical codes are necessary across public and private sectors to maintain order and trust in public service delivery.

Misconduct in governance, which results in a loss of public trust and resources, highlights the importance of ethics in public service. This global and national concern drives efforts to restore integrity in public institutions, safeguard democracy and promote better governance. Ethical codes also ensure political neutrality by maintaining a clear separation between the roles of public servants and elected representatives, preventing conflicts of interest. Furthermore, the changing role of the State in response to globalization, technological advancements and other modern challenges necessitates a reformed ethical framework to address new issues like crime and intellectual property rights.

Adherence to ethical codes fosters accountability, ensuring that public services are delivered fairly and promoting trust between the public and the administration. These codes also set standards for professionalism, guiding public sector employees in their duties and interactions with the public. Transparency in ethical codes enhances accountability, improving public trust in government institutions. Moreover, respect for individuals is a fundamental aspect of ethical codes, promoting fairness and dignity in interactions between the government and the public, which is crucial for a safe and healthy work environment.An OECD Study Group suggested ten principles for ethical conduct in public administration, which can be adapted to suit country-specific needs. These principles include clarity in ethical standards, legal enforcement and the availability of ethical guidance to public servants. Public officials must know their rights and obligations, supported by clear rules and a robust legal framework. Political leaders must demonstrate commitment to ethical behaviour, ensuring transparency and scrutiny in decision-making processes. Ethical codes should also guide the interaction between the public and private sectors, promoting integrity in these relationships. Public service conditions and human resource management policies should systematically encourage ethical behaviour. Accountability mechanisms must be in place at all levels to ensure ethical conduct and appropriate procedures should be established to address misconduct effectively.These principles help maintain ethical conduct among public servants, fostering an environment where the public interest is served with integrity, transparency and fairness. When effectively implemented, ethical codes can ensure that public servants' actions are aligned with the broader goals of good governance and public trust.

Suggested Readings

Avolio, B. J., & Gardner, W. L. (2005). Authentic leadership development: Getting to the root of positive forms of leadership. *The Leadership Quarterly, 16*(3), 315-338.

Bass, B. M. (1985). Leadership and performance beyond expectations. Free Press.

Ciulla, J. B. (2014). Ethics, the heart of leadership (3rd ed.). Praeger.

De Colle, S., Henriques, A., & Sarasvathy, S. (2014). The Paradox of Corporate Social Responsibility Standards. *Journal of Business Ethics*, 125(2), 177-191.

Greenleaf, R. K. (1977). Servant Leadership: A Journey into the Nature of Legitimate Power and Greatness. Paulist Press.

International Organization for Standardization (ISO). (2010). ISO 26000:2010 - Guidance on social responsibility. ISO.

KPMG. (2020). *The Time Has Come: The KPMG Survey of Sustainability Reporting 2020*. Retrieved from https://home.kpmg/xx/en/home/insights/2020/11/the-time -has-come-survey-of-sustainability-reporting.html (Accessed on 10 August 2024).

Porter, M. E., & Kramer, M. R. (2006). Strategy and society: The link between competitive advantage and corporate social responsibility. *Harvard Business Review, 84*(12), 78-92.

Robinson, Mark, From Old Public Administration to the New Public Service– Implications or Public Sector Reform in Developing Countries, (Singapore: Global Centre for Public Service Excellence, 2015).

Unilever. (2023). *Sustainable Living*. Retrieved from https://www.unilever.com /planet-and-society/sustainable-living/ (Accessed on 10 August 2024).

World Business Council for Sustainable Development. (1999). *Corporate Social Responsibility: Meeting Changing Expectations*. World Business Council for Sustainable Development.

Adapted from chromeextension://efaidnbmnnnibpcajpcglclefindmkaj/https:// one.oecd.org/document/C(98)70/FINAL/en/pdf (accessed on 12 February 2018). (Accessed on 10 August 2024).

CHAPTER 11

Navigating Sustainable Development

Challenges, Evaluation Metrics and Future Directions

As we reach the concluding chapter of this comprehensive exploration of sustainable development, it is essential to consolidate the discussions from previous chapters and chart the course for the future. Throughout this book, we have delved into the evolution of development administration from traditional paradigms to the modern-day focus on sustainability. We have examined various theories and models that underpin development administration, observed the shortcomings that necessitated a paradigm shift toward sustainable development and considered how these concepts are put into practice through diverse models and case studies. International agencies are pivotal in facilitating global sustainable development, underscoring the importance of citizen-centric governance, inclusive development and gender equality. Environmental concerns are an integral part of this discourse, as is the ethical dimension, which raises critical questions about governance and responsibility.

This final chapter synthesises these themes by addressing the contemporary challenges of sustainable development. It evaluates the effectiveness of existing metrics and offers insights into future directions that could help overcome these challenges. The complexity of sustainable development lies in its multifaceted nature – encompassing environmental, economic, social and ethical dimensions. Each aspect presents unique challenges that must be navigated with care and foresight. For instance, environmental sustainability often conflicts with economic growth, creating a dilemma for policymakers. Social equity, another crucial component, is frequently compromised when pursuing economic efficiency. The institutional and governance frameworks necessary to support sustainable development are often fragmented, leading to inconsistent and inadequate implementation of policies.

It is crucial to employ robust evaluation metrics that can accurately measure progress across various dimensions of sustainability to navigate these challenges. The development and adoption of these metrics have been discussed extensively, emphasizing the need for integrated approaches that combine qualitative and quantitative data. The role of technology, innovation and global governance in shaping the future of sustainable development is also highlighted, offering a glimpse into the possibilities that lie ahead. Moreover, the chapter underscores the importance of education and public awareness in fostering an environment of sustainability, which is indispensable for long-term success.

Thus, the path forward in sustainable development requires a concerted effort from all stakeholders – governments, businesses, civil society and individuals. It demands a commitment to innovation, collaboration and ethical governance. As we move towards a future where sustainability is not just an ideal but a necessity, the lessons learned from past successes and failures will be instrumental in guiding our steps. This chapter serves not only as a conclusion to the discussions in this book but also as a call to action, urging us all to engage actively in the pursuit of sustainable development for the benefit of current and future generations.

Introduction

Sustainable development is a concept that has evolved, shaped by various economic, social and environmental challenges. The roots of sustainable development can be traced back to the early environmental movements of the twentieth century, particularly in response to the Industrial Revolution and its impact on natural resources. The idea gained prominence in the 1980s with the publication of the Brundtland Report, formally known as 'Our Common Future', which defined sustainable development as 'development that meets the needs of the present without compromising the ability of future generations to meet their own needs'.[1] This definition highlighted the need for an integrated approach that balances economic growth, environmental protection and social equity.

Since then, sustainable development has become a guiding principle for global policy, influencing international agreements such as the Rio Earth Summit (1992), the Kyoto Protocol (1997) and the Paris Agreement (2015). The adoption of the United Nations' Sustainable Development Goals (SDGs) in 2015 further solidified sustainable development as a global priority, setting 17 goals and 169 targets to be achieved by 2030.[2] These goals cater to a wide range of areas, including poverty, hunger, health, education, gender equality, clean water and climate action, reflecting the interconnected nature of sustainable development challenges. However, before diving into the challenges in sustainable development, one must understand the failure of MDGs that led to the establishment of SDGs.

1 World Commission on Environment and Development, *Our Common Future* (Oxford: Oxford University Press, 1987).

2 United Nations, *Transforming Our World: The 2030 Agenda for Sustainable Development* (New York: United Nations, 2015).

Failure of Millennium Development Goals (MDGS)

Most of the underdeveloped countries, especially Africa, missed the MDGs by a large margin. Whatever progress occurred happened in a very uneven manner leaving most of the poor countries and especially children behind. MDGs were presented as a list of common goals for the global community to be achieved by 2015. However, the Civil Society galvanised into working towards the achievement of the goals. It has been observed by many scholars that probably the MDGs had been conceived as unrealistic and overambitious expectations which led to disillusionment and eroded the prospects of long-term engagement with the developing world. Almost 15.5 per cent of the world population still lives in poverty and suffers from hunger.

Following are some of the areas in which the MDGs faced failures according to the United Nations' final report.

1. Poverty: The UN goal was a 50 per cent reduction in the percentage of people living on under twenty-five dollars a day, but in South Asia, it was at 17 per cent and it remained at forty 1 per cent in sub-Saharan Africa.
2. Maternal mortality: The maternal mortality ratio of women aged 15–49 had been cut by half since 1990 but it was significantly short by 25 per cent from the goal. Out of 100,000 women, 510 in Sub-Saharan Africa are still dying from birthing every year, over twice as much as any other region in the world.
3. Focus on inputs: Goal 2 of the MDGs was to 'achieve universal primary education' which aimed at measuring enrolment rates and many poor countries made significant progress towards the goal, but whether they learnt anything remains a question. There was a rapid expansion of schools but it was found that there was a deteriorating effect on the learning quality due to shortages of teachers, resulting in single-teacher schools responsible for one multigrade classroom. One hundred and thirty million children completed primary education without being able to read and write. Hence, they focused more on inputs rather than focusing on the outputs and improving the quality of lives of the people.
4. MDGs as a tool for donors: Although the MDGs were signed by a huge number of developing countries, they had not participated in framing and writing them. The donor agencies had themselves identified the areas that they wanted to raise funds and donate for instead of concentrating on the problems the developing countries had. One example of this peculiar problem is of Rwanda where in 2014, 24 per cent of its health spending was on HIV/AIDS, although only 1.6 per cent of its population was affected by it.

5. Global emissions: The numbers for global emissions of carbon dioxide and water scarcity are disheartening. There has been a 50 per cent increase in CO_2 emissions and water scarcity which affects more than 40 per cent of the world in comparison to the 1990 statistics.
6. Goal 8 on developing a global partnership for development: The only MDG directed specifically at high-income countries has been disappointing. As a possible result of the global financial crisis, development aid has fallen for the first time in more than a decade. In instances where MDGs have been achieved, some of these successes are considered as a 'by-product of the rapid economic growth of China and India' rather than achievements of MDG-oriented activities.

However, the MDGs catapulted the world into thinking about the issues of development and the urgent need to do something which led to the formation of the SDG 2030 Agenda.

Challenges in Sustainable Development

Although sustainable development is discussed in all circles, significant challenges remain in its implementation. These challenges can be broadly categorised into environmental, economic, social, and institutional dimensions given below:

Environmental challenges in sustainable development

Environmental challenges are perhaps the most visible and immediate concerns within the sustainable development framework. Among these, climate change stands out as the most pressing global issue. As outlined by the Intergovernmental Panel on Climate Change (IPCC), the scientific consensus is that human tasks, especially burning fossil fuels, are the central drivers of global warming.[3]

The consequences of climate change are far-reaching, affecting weather patterns, sea levels and biodiversity, and presenting considerable risks to human health, food security and infrastructure. For example, sea levels rise due to the thawing of polar ice caps and glaciers, threatening coastal communities and ecosystems. According to the IPCC's 2021 report, the levels of sea worldwide could rise up to one meter by 2100 if current trends continue, which would have devastating impacts on millions of people living in low-lying areas. More recurrent and intense weather events, such as hurricanes, droughts, and floods,

3 Intergovernmental Panel on Climate Change, Climate Change 2021: The Physical Science Basis, *Contribution of Working Group I to the Sixth Assessment Report of the Intergovernmental Panel on Climate Change* (Cambridge: Cambridge University Press, 2021).

are already being observed, leading to loss of life, displacement of populations, and economic damages that disproportionately affect developing countries.

Biodiversity loss is another critical environmental challenge. The Earth is currently undergoing a significant species decline, commonly known as the sixth mass extinction, driven by habitat destruction, climate change, pollution and overexploitation of natural resources.[4] Biodiversity is essential for the resilience of ecosystems, providing services such as pollination, nutrient cycling, and water purification that are vital for human survival. The loss of biodiversity also undermines ecosystems' cultural and economic value, particularly for indigenous communities who rely on biodiversity for their livelihoods.

Pollution presents another formidable environmental challenge, particularly air, water and soil pollution. The World Health Organization (WHO) estimates that air pollution alone is responsible for around seven million premature deaths annually, with the majority occurring in low- and middle-income countries.[5] Air pollution, caused by emissions from vehicles, industries and agricultural activities, not only impacts human health but also plays a role in driving climate change through the emission of greenhouse gases and black carbon. Water pollution, on the other hand, poses a threat to both aquatic ecosystems and human populations. Contaminants such as plastics, heavy metals and pesticides enter water bodies through agricultural runoff, industrial discharge and improper waste disposal, leading to the degradation of freshwater and marine environments.[6] This pollution has severe consequences for marine life, including coral reefs and fish populations, leading to the pollution of drinking water supplies and posing health risks to millions of people worldwide. Soil pollution, often overlooked, is equally concerning.

The use of pesticides, fertilisers and industrial chemicals has led to the degradation of soils, reducing their fertility and ability to support agriculture.[7] This, in turn, threatens food security, particularly in regions where agriculture is the

4 G. Ceballos et al., 'Biological annihilation via the ongoing sixth mass extinction signalled by vertebrate population losses and declines', *Proceedings of the National Academy of Sciences*, 114.30 (2017), E6089–E6096.

5 World Health Organization, *World Health Statistics 2016: Monitoring Health for the SDGs* (Geneva: World Health Organization, 2016).

6 UNESCO, *Global Education Monitoring Report 2019: Migration, Displacement and Education – Building Bridges, Not Walls* (United Nations Educational, Scientific and Cultural Organization, 2019), available at: <https://en.unesco.org/gem-report/report/2019/migration-displacement-and-education>, accessed 6 September 2024.

7 FAO, *The State of the World's Forests 2018: Forest Pathways to Sustainable Development* (Food and Agriculture Organization of the United Nations, 2018), available at: <http://www.fao.org/state-of-forests/en>, accessed 6 September 2024.

primary livelihood. Contaminated soils can accumulate toxic substances in the food chain, posing additional risks to human health. Sadhguru, a spiritual leader and an environmentalist, has begun a campaign for soil conservation, 'Rally for Rivers', to address soil degradation and water scarcity. It promotes sustainable agricultural practices and the restoration of natural water bodies to rejuvenate soil health, increase water retention and support biodiversity, emphasizing the critical role of healthy soil for environmental and human well-being.

Dealing with these environmental challenges requires a multifaceted approach that includes robust regulatory frameworks, technological innovation and changes in consumer behaviour. Global treaties, like the Paris Agreement, are essential in uniting international efforts to address climate change and advance sustainability initiatives. However, effective implementation at the national and local levels is vital to ensure that these commitments translate into real-world action. Technological innovation also offers potential solutions to environmental challenges. Renewable energy technologies like solar and wind power are critical for reducing greenhouse gas emissions and transitioning to a low-carbon economy.[8] Advances in energy efficiency, sustainable agriculture and waste management can also help mitigate environmental impacts and promote sustainable development. Finally, changes in consumer behaviour, supported by education and awareness campaigns, are vital for reducing the environmental footprint of human activities. Shifting towards more sustainable consumption patterns, such as reducing meat consumption, minimizing waste and choosing eco-friendly products, can significantly reduce the demand for resources and the associated environmental impacts.

Hence, the environmental challenges facing sustainable development are complex and interconnected, requiring a comprehensive and coordinated response. Addressing these challenges is essential for protecting the planet and ensuring the well-being of current and future generations.

Economic challenges in sustainable development

Economic challenges are deeply embedded within the broader context of sustainable development, intersecting with environmental and social dimensions in complex ways. The traditional economic growth model, measured by Gross Domestic Product (GDP), has long been criticised for its narrow focus on material wealth and failure to account for environmental degradation and social

8 IRENA, *Future of Wind: Deployment, Investment, Technology, Grid Integration and Socio-Economic Aspects* (International Renewable Energy Agency, 2019), available at: <https://www.irena.org/publications/2019/Oct/Future-of-wind>, accessed 6 September 2024.

inequalities.[9] While GDP growth has been associated with improved living standards, it often overlooks the depletion of natural resources and the long-term sustainability of economic activities.

One of the critical economic challenges is the inherent tension between economic growth and environmental sustainability. This tension is exemplified by the 'growth versus sustainability' dilemma that occurs when pursuing economic expansion, which leads to environmental degradation. For example, the rapid industrialization of emerging economies, while driving economic growth, has resulted in severe air and water pollution, deforestation and biodiversity loss.[10] The environmental impacts of this growth are frequently shifted outside the market, meaning they are not included in the prices of goods and services, which results in overconsumption and unsustainable use of resources.

Ways to tackle economic challenges

Economists and policymakers have proposed alternative metrics for measuring progress that extend beyond GDP to address this challenge. They are:

5. Genuine Progress Indicator: One such measure is the Genuine Progress Indicator (GPI), which adjusts GDP by considering environmental degradation, social well-being and income inequality.[11] The GPI aims to provide a more holistic view of economic progress by considering the sustainability of economic activities and their impact on human welfare. For example, 'The GPI: An Alternative to GDP' by various environmental and economic research institutes and State-Level GPI Reports in the U.S., such as those conducted in Maryland or Vermont.
   - Circular economy: Another approach to resolving the growth versus sustainability dilemma is the concept of a 'circular economy'. The circular economy model seeks to decouple economic growth from resource consumption by promoting reuse, recycling, and regeneration of materials. This model departs from the conventional linear economy, where

9 J. E. Stiglitz, A. Sen, and J.-P. Fitoussi, *Report by the Commission on the Measurement of Economic Performance and Social Progress* (Commission on the Measurement of Economic Performance and Social Progress, 2009), available at: <https://www.stiglitz-sen-fitoussi.fr/en/index.htm>, accessed 6 September 2024.

10 T. Panayotou, *Economic Growth and the Environment* (Harvard Institute for International Development, 1993), available at: <https://www.hiid.harvard.edu>, accessed 6 September 2024.

11 I. Kubiszewski et al., 'The genuine progress indicator 1950–2009: Methodology and results', *Ecological Economics*, 93 (2013), 41–58, available at: <https://doi.org/10.1016/j.ecolecon.2013.06.009>, accessed 6 September 2024.

resources are extracted, used, and disposed of as waste. The circular economy reduces environmental impacts and creates economic opportunities in areas such as recycling, remanufacturing, and sustainable product design by closing the loop of resource use. The transition to a circular economy requires significant changes in production and consumption patterns and supportive policies and regulations. For instance, the European Union has adopted a Circular Economy Action Plan to foster sustainable growth and reduce waste by encouraging product design that prioritises durability, repairability, and recyclability.[12] Implementing the circular economy globally poses challenges, particularly in harmonizing regulations across countries and ensuring the benefits are equitably distributed.

- Global inequality: Global inequality is another significant economic challenge in the context of sustainable development. The unequal distribution of wealth and resources within and between countries exacerbates environmental and social problems. For instance, the wealthiest 10 per cent of the global population is responsible for nearly half of global carbon emissions, while the poorest 50 per cent contribute only 10 per cent.[13] This disparity highlights the need for environmental sustainability and social equity policies. The challenge of global inequality is closely linked to poverty, which remains widespread in many parts of the world. Poverty limits access to primary resources such as clean water, food, and energy, making it difficult for individuals and communities to pursue sustainable development. Moreover, people with low incomes are often the most vulnerable to environmental changes, such as climate change and natural disasters, as they have fewer resources to adapt or recover from such events.[14] Recent reports highlight a significant and growing problem of global inequality, exacerbated by multiple crises like the pandemic, climate change, and economic instability. Oxfam's 2023 report, *Survival of the Richest*, reveals that the wealthiest billionaires saw a surge in wealth while poverty rates and the cost of living worsened globally.

---

12 European Commission, *The European Green Deal* (European Commission, 2020), available at: <https://ec.europa.eu/info/publications/communication-european-green-deal_en>, accessed 6 September 2024.

13 Oxfam, *Time to Care: Unpaid and Undervalued Care Work and the Global Inequality Crisis* (Oxfam International, 2020), available at: <https://oxfamapps.org/media/press_releases/oxfam-time-to-care-report/>, accessed 6 September 2024.

14 World Bank, *Poverty and Shared Prosperity 2020: Reversals of Fortune* (The World Bank Group, 2020), available at: <https://www.worldbank.org/en/publication/poverty-and-shared-prosperity>, accessed 6 September 2024.

> The wealthiest 10 per cent now control 52 per cent of global income, while the poorest 50 per cent earn just 8.5 per cent.

Sustainable development efforts must focus on creating inclusive economic opportunities that benefit all segments of society to deal with inequalities. This includes investing in education, healthcare, and infrastructure, particularly in developing countries, to enable people to lift themselves out of poverty and contribute to sustainable development. Additionally, policies that promote fair trade, labour rights and corporate social responsibility are essential for ensuring that economic activities are conducted to respect human rights and the environment. Thus, the economic challenges of sustainable development are multifaceted and require a shift away from traditional growth paradigms towards more sustainable and inclusive models. The international community can work towards a future where economic activities support rather than undermine sustainable development by adopting alternative measures of progress, promoting circular economy principles, and addressing global inequality.

5. Social Challenges in Sustainable Development: Social challenges are at the heart of sustainable development, encompassing issues of equity, inclusion, and social justice. These challenges are deeply interconnected with environmental and economic dimensions, influencing how societies use natural resources and generate financial. Social challenges must be recognised and resolved to achieve sustainable development that benefits all individuals and communities, particularly the marginalised or vulnerable. They are:
    - Global poverty: One of the most pressing social challenges is global poverty, a significant barrier to sustainable development. Despite progress in reducing poverty levels over the past few decades, approximately 9.2% of the world's population still lives on less than $1.90 a day, the international poverty line.[15] Poverty involves more than just insufficient income; it also encompasses inadequate access to essential services like education, healthcare, clean water, and sanitation.
    - Impact: Environmental degradation and climate change adversely impact people experiencing poverty as they rely heavily on natural resources for their livelihoods and cannot adapt to changes.
    - Inequality: Inequality is a major social challenge in sustainable development. Economic inequality within and between countries has risen in

15 D. Dorling, *Inequality and the 1%* (Verso Books, 2019)

recent decades, leading to disparities in income, wealth, and access to opportunities.[16] Inequality undermines social cohesion and stability, as it can lead to social unrest, crime, and political instability. Moreover, inequality can exacerbate environmental degradation, as wealthier individuals and countries tend to have larger ecological footprints, consuming more resources and generating more waste.[17]

- Social exclusion: Social exclusion is another critical issue, as it prevents certain groups from fully participating in economic, social, and political life. Exclusion can be based on various factors, including race, ethnicity, gender, age, disability, and sexual orientation. For example, women and girls often face discrimination and violence, limiting their access to education, healthcare, and employment opportunities.[18] Social exclusion not only violates human rights but also hinders sustainable development by preventing the full potential of all individuals from being realised.

Ways to overcome challenges in sustainable development

Social challenges in sustainable development require a focus on inclusive development to ensure that all individuals and communities benefit from development efforts. Inclusive development involves creating opportunities for marginalised and vulnerable groups, reducing inequalities, and promoting social justice. This can be accomplished through policies and programs that tackle the root causes of poverty and inequality, such as investing in education, healthcare, and social protection systems, briefly described here.[19]

- Education: Education plays a crucial role in advancing sustainable development by equipping individuals with the knowledge and skills necessary to engage effectively in economic, social and political activities.
- Education: Education also raises awareness of environmental issues and promotes sustainable behaviours, such as energy conservation and waste

---

16 World Bank, *Poverty and Shared Prosperity 2021: Reversals of Fortune* (The World Bank Group, 2021), available at: https://www.worldbank.org/en/publication/poverty-and-shared-prosperity [accessed 6 September 2024].

17 T. Piketty, *Capital in the Twenty-First Century* (Harvard University Press, 2014).

18 UN Women, *Gender Equality: Women's Rights in Review 25 Years after Beijing* (New York: UN Women, 2020).

19 UNESCO, *Education for Sustainable Development Goals: Learning Objectives* (United Nations Educational, Scientific and Cultural Organization, 2017)

reduction.[20] Ensuring access to quality education for all, particularly girls and marginalised groups, is essential for sustainable development.

- Healthcare: Healthcare is another critical area, as good health is fundamental to human well-being and development. Access to healthcare services, including maternal and child health, nutrition and disease prevention, is essential for reducing poverty and inequality.[21] The COVID-19 pandemic has highlighted the importance of strong healthcare systems in addressing global health challenges and ensuring that all individuals have the opportunity to lead healthy lives.
- Social welfare: Social protection systems, such as unemployment benefits, pensions and social safety nets, are also critical for reducing poverty and inequality and promoting social inclusion. These systems provide a buffer against economic shocks and help individuals and families cope with the risks and uncertainties of life.[22] Expanding social protection coverage, particularly in developing countries, is essential for sustainable development.

Hence, social challenges are central to the sustainable development agenda, as they affect the well-being and opportunities of individuals and communities. Addressing these challenges requires a focus on inclusive development, which promotes equity, social justice, and the participation of all individuals in society. Investing in education, healthcare, and social protection and reducing inequalities and social exclusion can ensure sustainable development that benefits everyone.

1. Institutional and policy challenges in sustainable development: The institutional and policy dimensions of sustainable development are critical for translating global goals into actionable strategies and practices. Effective governance, robust institutions and coherent policies are necessary to deal with complex and interconnected challenges for sustainable development. However, the current global governance system faces significant challenges in ensuring that sustainable development is equitable, effective, and inclusive. They are:

20 WHO (World Health Organization), *Air Pollution* (Geneva: World Health Organization, 2019), available at: <https://www.who.int/health-topics/air-pollution>, accessed 1 September 2024.

21 UN Women, *Gender Equality: Women's Rights in Review 25 Years after Beijing* (New York: UN Women, 2020).

22 UNDP, *Network Governance for Sustainable Development* (New York: United Nations Development Programme, 2016).

- Governance fragmentation: One of the significant institutional challenges is the fragmentation of governance structures at the international and national levels. The global governance system for sustainable development is characterised by a multitude of institutions, agreements, and initiatives, each with its mandate and agenda. While this diversity can be a strength, it also leads to coordination problems, overlap and inefficiencies. For example, the multitude of international environmental agreements, such as the Convention on Biological Diversity (CBD) and the United Nations Framework Convention on Climate Change (UNFCCC), often operate in silos, making it challenging to address the interrelated nature of environmental issues.[23]
- Weak institutions: At the national level, implementing sustainable development policies faces challenges due to weak institutions, lack of capacity and political instability. Governments struggle to implement and enforce environmental and social policies in many developing countries due to limited financial and human resources, corruption, and competing priorities. This undermines the effectiveness of sustainable development efforts and can lead to adverse outcomes, such as environmental degradation, social exclusion, and economic inequality.
- Policy integration: Policy coherence is another significant challenge. Sustainable development involves the integration of environmental, economic and social objectives into policymaking, often balancing competing interests and trade-offs. However, policies are frequently developed in isolation, leading to conflicts and unintended consequences. For example, policies promoting economic growth through industrialization can lead to environmental degradation without measures to protect natural resources and reduce pollution.[23]
- Interest groups: The policymaking process is often influenced by powerful interest groups, such as large corporations and industries, which can lead to policies prioritizing short-term economic gains over long-term sustainability.[24] This can result in adopting policies detrimental to the environment and social well-being, such as subsidies for fossil fuels or deforestation for agricultural expansion.
- Global collaboration: International cooperation is essential for addressing the global nature of sustainable development challenges. However,

---

23 K. Schwab, *The Fourth Industrial Revolution* (New York: Crown Business, 2017).

24 IPCC (Intergovernmental Panel on Climate Change), *Global Warming of 1.5°C: An IPCC Special Report (2018)*, available at: <https://www.ipcc.ch/sr15/>, accessed 1 September 2024.

achieving consensus among countries with diverse interests, priorities, and levels of development is difficult. For instance, the negotiations under the UNFCCC have been characterised by disagreements between developed and developing countries over issues such as responsibility for greenhouse gas emissions, financial support, and technology transfer. These disagreements have slowed progress in reaching binding agreements and implementing effective climate action.

Improving institutional and policy challenges

Several approaches can be adopted to strengthen policies and institutions. Strengthening global governance for sustainable development requires greater coordination among international institutions, improved policy coherence and more inclusive decision-making processes. Creating stronger links between existing institutions, such as the UN Environment Programme (UNEP) and the World Bank, is critical to ensure that environmental, economic, and social issues are addressed in an integrated manner. Some ways are:

- Institutional capacity: At the national level, building institutional capacity is essential for effective policy implementation. Strengthening the rule of law, improving transparency and accountability, and increasing financial and technical support for developing countries will result in more substantial institutional capacity. Enhancing public participation in decision-making processes is also essential, as it ensures that policies are responsive to the needs and priorities of all stakeholders, including marginalised and vulnerable groups.[25]
- Enhancing policy alignment for sustainable development: Improving policy coherence involves adopting integrated approaches to policymaking that consider the interconnections between environmental, economic, and social objectives. Other vital steps toward this goal include breaking down silos within government agencies, fostering collaboration across sectors and levels of government and adopting long-term planning frameworks that prioritise sustainability and resilience over short-term economic gains.

  Building trust: International cooperation can be enhanced by building trust among countries, particularly between developed and developing nations. Building trust requires recognizing countries' common but differentiated responsibilities in dealing with global challenges. For example, developed

25 K. Raworth, *Doughnut Economics: Seven Ways to Think Like a 21st-Century Economist* (Chelsea Green Publishing, 2017).

countries agreed to take the lead in reducing greenhouse gas emissions in the Paris Agreement on climate change. In contrast, developing countries with fewer resources were allowed more flexibility in their commitments. This approach recognises the varying capacities and historical contributions to environmental degradation, helping build trust between nations by ensuring that responsibilities are shared equitably.

- Providing financial and technical support to developing countries is essential for building trust. International organizations like the UN play a key role in fostering dialogue and cooperation, which are crucial for global sustainable development.

So, sustainable development's institutional and policy challenges are significant, but not insurmountable. Strengthening governance structures, building institutional capacity, improving policy coherence, and enhancing international cooperation are critical steps for overcoming these challenges. The global community can ensure that sustainable development is achieved effectively, equitably, and inclusively through these efforts.

Monitoring sustainable development governance

Monitoring is a crucial element in the framework of sustainable development governance. It involves the continuous observation and measurement of progress towards achieving sustainability goals, ensuring that policies and initiatives are effective, and identifying areas where adjustments may be needed. Effective monitoring systems provide the necessary data to inform decision-making, foster accountability, and drive improvements in governance practices. Monitoring entails gathering, examining and reporting data concerning sustainability indicators. These indicators can be quantitative, such as emissions levels, energy consumption, or economic growth rates, or qualitative, such as stakeholder satisfaction or policy implementation quality. By tracking these indicators, governments and organizations can assess how well their strategies are performing and whether they are on track to meet their sustainability targets.

One of the primary functions of monitoring is to provide real-time feedback on the effectiveness of policies and programs. This feedback is crucial for adaptive management, which allows modifying strategies based on performance data. For instance, if monitoring reveals that a particular environmental regulation is not leading to the expected reductions in pollution, policymakers can use this information to revise the regulation, implement stricter standards, or explore alternative approaches. This iterative process helps ensure that governance systems remain responsive and effective in addressing emerging challenges.

Monitoring also plays a significant role in ensuring transparency and accountability in sustainable development governance. Governments and organizations can show their dedication to sustainability and allow stakeholders, including the general public, civil society and private sector entities, to scrutinise their actions by publicly monitoring data. This openness helps build trust and encourages greater stakeholder engagement in sustainability efforts. Transparency in monitoring also guarantees that governance processes are carried out fairly and accountably, which helps minimise the risk of corruption and mismanagement. The selection of appropriate indicators is a critical aspect of effective monitoring. Indicators must be relevant, measurable, and capable of reflecting the dimensions of sustainable development. For example, environmental monitoring might include air and water quality indicators, biodiversity levels, and resource use efficiency. Social indicators might focus on poverty reduction, education levels, and health outcomes. Economic indicators could encompass GDP growth, employment rates and income distribution. The choice of indicators should align with the specific goals and context of the sustainability initiatives being monitored.

Challenges in monitoring sustainable development governance include data availability, accuracy and the complexity of integrating diverse data types. Many regions, especially developing countries, may have limited data collection and analysis capacity. This can hinder the ability to conduct effective monitoring and to make informed decisions based on accurate information.[26] Investment in data collection infrastructure, personnel training and innovative technologies like remote sensing and data analytics is necessary to improve monitoring capabilities. Monitoring systems must also be designed to account for the dynamic and interconnected nature of sustainable development issues. Sustainable development encompasses multiple dimensions – environmental, social and economic – that interact in complex ways. Effective monitoring requires a holistic approach that considers these interactions and assesses the overall impact of policies and initiatives across different areas. For example, a policy aimed at economic development might have environmental sustainability and social equity implications, necessitating a comprehensive monitoring framework that captures these multidimensional effects.[29]

Hence, monitoring is an essential component of sustainable development governance, providing the data and insights needed to evaluate the effectiveness of policies and programs, ensure transparency and accountability, and support adaptive management. By selecting relevant indicators, addressing data

---

26 OECD (Organisation for Economic Cooperation and Development), *OECD Business and Finance Outlook 2020: Sustainable and Resilient Finance* (Paris: OECD Publishing, 2020.

challenges, and adopting a holistic approach, monitoring systems can enhance the effectiveness of governance practices and contribute to achieving sustainability goals. As the global community continues to address pressing sustainability challenges, robust and effective monitoring will be key to driving progress and ensuring that governance systems are responsive and accountable.

Evaluation metrics for sustainable development

Evaluation metrics are essential tools for tracking progress towards sustainable development goals (SDGs) and ensuring accountability in their implementation. These metrics provide a way to measure the effectiveness of policies and initiatives, identify gaps and challenges and guide decision-making processes. However, developing and applying evaluation metrics for sustainable development is a complex task, given the multidimensional nature of sustainability and the diversity of contexts in which it is pursued.

One of the primary challenges in evaluating sustainable development is the need for comprehensive and integrated metrics that capture its environmental, economic and social dimensions. Traditional metrics, such as GDP, are inadequate for this purpose, as they focus primarily on economic output and fail to account for environmental degradation and social inequalities. As a result, alternative metrics have been developed to provide a more holistic view of progress.

1. SDGs: One of the most widely used sets of metrics for sustainable development is the Sustainable Development Goals (SDGs) indicators. The SDGs, adopted by the United Nations in 2015, consist of 17 goals and 169 targets that cover a broad range of sustainability issues, including poverty, inequality, health, education, clean water and climate action.[30] Each target is accompanied by one or more indicators used to track progress and assess the effectiveness of policies and interventions. The SDG indicators represent a significant advancement in the measurement of sustainable development, as they provide a comprehensive framework that applies to all countries, regardless of their level of development. However, implementing these indicators faces challenges, particularly regarding data availability, quality, and comparability. Many developing countries lack the statistical capacity to collect and analyse the data required for the SDG indicators, which can hinder their ability to monitor progress and identify areas for improvement.
2. Human Development Index: Other composite indices besides the SDG indicators have been developed to evaluate sustainable development. One example is the Human Development Index (HDI), which measures human development based on life expectancy, education and income. The HDI

provides a more comprehensive view of development than GDP, considering economic and social factors. However, the HDI has been criticised for its limited income focus and failure to account for environmental sustainability.

3. The Environmental Performance Index: The Environmental Performance Index (EPI) is another composite index focusing on environmental sustainability. The EPI, developed by Yale University and Columbia University, ranks countries based on their performance in air quality, water and sanitation, biodiversity and habitat, and climate and energy.[31] The EPI provides valuable insights into countries' environmental performance and helps identify best practices and areas for improvement. However, like other composite indices, the EPI has limitations, particularly regarding data availability and the selection of indicators.

A growing interest exists in developing more context-specific and participatory evaluation approaches to address existing metrics' limitations. Context-specific metrics are tailored to a particular region or community's unique social, environmental, and economic conditions, allowing for a more accurate assessment of progress towards sustainability.[32] Participatory approaches, in contrast, involve stakeholders in developing and applying metrics, ensuring that they reflect local priorities and values. For example, using community-based indicators in rural development projects allows local communities to define their indicators of success based on their specific needs and aspirations.[33] This approach enhances the relevance and legitimacy of evaluation metrics and empowers communities to take ownership of the development process.

Thus, evaluation metrics are essential for tracking progress towards sustainable development and ensuring accountability in implementing sustainability goals. While existing metrics, such as the SDG indicators, the HDI, and the EPI, provide valuable insights, they also have limitations that must be addressed. Developing more comprehensive, context-specific, and participatory approaches to evaluation can help overcome these challenges and ensure that sustainable development is measured in a way that is accurate, inclusive and relevant to the needs of all stakeholders.

The need for collaborative and network governance in promoting sustainable development governance

Sustainable development governance is a complex and multifaceted challenge that requires the coordination and integration of diverse actors, sectors and levels of governance. The interconnected nature of environmental, social and economic issues demands approaches beyond traditional hierarchical governance models. In this context, collaborative and network governance has emerged as a critical approach to effectively confront the diverse challenges of sustainable development. This approach emphasises the importance of cooperation, partnerships and decentralised decision-making, fostering more inclusive, resilient and adaptive governance systems.[34]

Collaborative governance involves the participation of multiple stakeholders, including governments, civil society organizations, businesses, academia and local communities, in the decision-making process. It is rooted in the recognition that no single actor or sector has the knowledge, resources, or capacity to address the complex challenges of sustainable development.[35] Collaborative governance enables co-creating more comprehensive and contextually relevant solutions by bringing together diverse perspectives and expertise. This approach also promotes transparency, accountability and shared responsibility, which are essential for building trust and ensuring the legitimacy of governance processes.[36]

Network governance, closely related to collaborative governance, refers to the horizontal and decentralised structures that connect various actors across different levels of administration. Unlike traditional top-down governance models, network governance is characterised by forming partnerships, alliances, and networks that facilitate information sharing, resource mobilization, and coordinated action.[37] These networks can be formal or informal and often transcend national borders, enabling global cooperation on climate change, biodiversity conservation, and poverty reduction.[38]

The need for collaborative and network governance in promoting sustainable development governance is particularly evident in global environmental challenges, such as climate change and biodiversity loss. These challenges are transboundary and require coordinated efforts at local, national, and international levels. For example, the Paris Agreement on climate change is a prime example of network governance, where multiple countries, cities, businesses, and non-governmental organizations (NGOs) work together to achieve shared climate goals. The success of such initiatives depends on the ability of diverse actors to collaborate, share knowledge and pool resources, demonstrating the critical role of network governance in addressing global challenges.[27]

Moreover, collaborative and network governance is essential for ensuring that sustainable development policies and practices are inclusive and equitable. These governance approaches can help ensure that the voices of marginalised and vulnerable groups are heard and that their needs are met by involving a

---

27 UNFCCC, *The Paris Agreement (2015)*, available at: <https://unfccc.int/process-and-meetings/the-paris-agreement/the-paris-agreement>, accessed 1 September 2024.

wide range of stakeholders in decision-making processes.[41] This inclusivity is vital for achieving social justice and equity, critical components of sustainable development. For instance, including Indigenous communities in environmental governance can lead to more effective conservation practices that respect traditional knowledge and cultural values.[42]

Another significant advantage of collaborative and network governance is their ability to foster innovation and adaptability in governance systems. In the face of rapidly changing environmental and social conditions, traditional governance structures often struggle to respond effectively. In contrast, networks of diverse actors can quickly adapt to new challenges, share innovative solutions, and implement them on time. This adaptability is crucial for achieving long-term sustainability in an era of uncertainty and change.[43] However, the implementation of collaborative and network governance is not without challenges. It requires overcoming institutional barriers, power imbalances and stakeholder differences in interests and values. Effective collaboration also demands strong leadership, effective communication and a commitment to building trust and mutual understanding. The success of network governance depends on managing complex interactions and maintaining the coherence and direction of collective efforts.

Hence, the need for collaborative and network governance to promote sustainable development governance is driven by the complexity and interconnectedness of the challenges we face. These governance approaches can help create more resilient and effective systems better equipped to achieve sustainable development goals by fostering cooperation, inclusivity and adaptability. As the world grapples with pressing environmental and social issues, the importance of collaborative and network governance will grow, making it an essential component of the global sustainability agenda.

Future directions for sustainable development

The sustainable development agenda faces opportunities and challenges as we look to the future. The next few decades will be critical in determining whether the global community can achieve the ambitious goals in the 2030 Agenda for Sustainable Development. This section explores some key trends and emerging issues that will shape the future of sustainable development, as well as the strategies and innovations that can help overcome these challenges.

Technological innovation and the fourth industrial revolution

The fourth industrial revolution, characterised by the convergence of digital, biological and physical technologies, has the potential to transform sustainable development. Technologies such as artificial intelligence (AI), blockchain, and the

Internet of Things (IoT) can enhance the efficiency and effectiveness of sustainability efforts by enabling smarter resource management, improving transparency and facilitating data-driven decision-making.[44] For example, AI can optimise energy use in buildings, reducing greenhouse gas emissions and lowering costs. Blockchain technology can improve supply chain transparency, ensuring products are sourced sustainably and ethically. The IoT can enable real-time monitoring of environmental conditions, helping to prevent and mitigate environmental risks.[45]

Advances in technology, particularly in areas such as artificial intelligence, biotechnology and renewable energy, have the potential to drive significant progress towards sustainability. For example, developing clean energy technologies, such as solar and wind power, is essential for reducing greenhouse gas emissions and transitioning to a low-carbon economy. Similarly, precision agriculture and biotechnology advances can help increase food production while minimizing environmental impacts. However, technological change also presents challenges, particularly in ensuring its benefits are equitably distributed. The digital divide, which refers to the gap between those with access to digital technologies and those without, is a significant barrier to inclusive development. Ensuring that all countries and communities can access and benefit from technological innovations is essential for sustainable development. There is the risk of increasing inequality, data privacy concerns, and the potential for unintended environmental consequences. Ensuring technological innovation aligns with sustainable development goals will require proactive governance, ethical considerations and inclusive policymaking.

Climate change and resilience

Climate change will continue to be one of the most significant challenges for sustainable development in the coming decades. The impacts of climate change, including more frequent and severe extreme weather events, rising sea levels, and shifting ecosystems, will have profound implications for food security, water availability, public health, and economic stability.[45] Resilience to climate change is crucial for sustainable development, requiring investments in climate-resilient infrastructure, sustainable agricultural practices and improved disaster risk management. Prioritizing climate justice is also increasingly important, as it seeks to address the disproportionate impacts of climate change on vulnerable communities and ensure that climate action is equitable and inclusive.

Systemic change

Another emerging issue is the growing recognition of the need for systemic change. The interconnected nature of global challenges, such as climate change, inequality and biodiversity loss, requires a systemic approach that addresses the root causes of these problems rather than just their symptoms. Rethinking current economic, social and political systems is essential for creating more sustainable and resilient societies. For example, the concept of the 'doughnut economy', proposed by economist Kate Raworth, offers a new framework for sustainable development that balances the needs of people and the planet.[46]

Sustainable finance and investment

The role of finance in sustainable development is increasingly recognised, with sustainable finance emerging as a paramount driver of the transition to a low-carbon and inclusive economy. Sustainable finance integrates environmental, social and governance (ESG) criteria into financial decision-making to promote long-term value creation and positive societal impact.[47] Green bonds, impact investing and ESG-linked financial products are examples of sustainable finance instruments gaining traction in global markets. These instruments can mobilise capital for sustainable projects like renewable energy, sustainable agriculture and affordable housing while driving corporate accountability and transparency.

However, the growth of sustainable finance also raises questions about greenwashing, the practice of making misleading claims about a product or investment's environmental or social benefits. Ensuring the credibility and integrity of sustainable finance will require robust standards, transparent reporting, and independent verification.[49] The role of finance in sustainable development is also gaining increasing attention. Mobilizing the necessary financial resources to achieve the SDGs is a significant challenge, particularly in the COVID-19 pandemic, which has strained public finances and increased debt levels in many countries.[49] Innovative financing mechanisms, such as green bonds, impact investing and blended finance, offer potential solutions for closing the financing gap and directing investments towards sustainable development projects.[50] The private sector will play a crucial role in the future of sustainable development as businesses increasingly recognise the importance of sustainability for long-term success. Corporate sustainability initiatives are becoming more widespread. All corporate organizations like to adopt environmental, social, and governance (ESG) criteria and publicise it to their advantage. This trend is driven by consumer demand, regulatory pressures, and a growing awareness of the financial risks associated with unsustainable practices. However, there is still a need for stringent accountability mechanisms to ensure that corporate sustainability commitments translate into meaningful action.

Social inclusion and equity

Social inclusion and equity will remain central to sustainable development, particularly in rising inequality, demographic shifts and social movements. Ensuring that development benefits all individuals and communities, particularly those who are marginalised or vulnerable, will be critical for achieving the SDGs. Inclusive development approaches that prioritise the needs and rights of disadvantaged groups, such as women, indigenous peoples and persons with disabilities, are essential for building more equitable and resilient societies. Promoting gender equality, protecting human rights, and ensuring access to critical services such as education, healthcare, and social protection are essential components.

Also, the COVID-19 pandemic has highlighted the importance of social protection systems in safeguarding livelihoods and reducing vulnerability. Strengthening social protection, particularly in low- and middle-income countries, will be essential for building resilience to future shocks and ensuring no one is left behind.[28] Changing consumption patterns, promoting sustainable lifestyles, and fostering a culture of sustainability are all critical components of the transition to a more sustainable future. Education and awareness-raising efforts will drive these changes, particularly among young people.

Global governance and multilateralism

The future of sustainable development will also depend on the strength of global governance and multilateralism. Addressing global challenges such as climate change, biodiversity loss and inequality requires coordinated international action and effective multilateral institutions. Nonetheless, recent years have seen growing challenges to multilateralism, including rising nationalism, geopolitical tensions and a retreat from international cooperation. Rebuilding trust in multilateral institutions, strengthening international norms, and promoting inclusive and participatory global governance will be critical for advancing sustainable development. The role of international organizations, such as the United Nations, World Bank, and International Monetary Fund, will continue to be important in providing leadership, setting standards, and mobilizing resources for sustainable development. However, non-state actors, including civil society, the private sector, and local communities, need greater involvement in global governance processes. Finally, the future of sustainable development will depend on the ability of the global community to work together in a spirit

---

28 International Labour Office (ILO), *World Social Protection Report 2017–19: Universal Social Protection to Achieve the Sustainable Development Goals* (International Labour Office, 2017).

of solidarity and cooperation. The COVID-19 pandemic has highlighted the importance of international cooperation in addressing shared challenges, but it has also exposed deep divisions and inequalities. Strengthening multilateralism and fostering a sense of global citisenship will be essential for overcoming these challenges and building a more sustainable and equitable world.

Thus, sustainable development's future is full of opportunities and challenges. Technological innovation, systemic change, innovative financing, and social and cultural shifts all have the potential to drive significant progress towards sustainability. However, realizing this potential will require strong leadership, global cooperation and a commitment to equity and justice. Embracing these principles allows the international community to navigate the complexities of sustainable development and create a better future for all.

Conclusion

The journey towards sustainable development is an ongoing process that requires continuous effort, innovation, and collaboration. Over the past few decades, significant progress has been made in raising awareness of sustainability issues and developing the frameworks, policies, and technologies needed to address them. However, the challenges remain formidable, and the path ahead is fraught with uncertainties and risks. This chapter has explored the various dimensions of sustainable development, including the environmental, economic, social and institutional challenges, as well as the evaluation metrics and future directions for sustainability. The discussion has highlighted the complexity and interconnectedness of these issues, underscoring the need for a holistic and integrated approach to sustainable development.

Environmental challenges, such as climate change, biodiversity loss and pollution, are among the most urgent threats to global sustainability. Robust regulatory frameworks, technological innovation, and significant changes in behaviour and consumption patterns are essential for effectively tackling these challenges. Economic challenges, including the tension between growth and sustainability, global inequality and poverty, further complicate the sustainability agenda. Achieving sustainable development will require rethinking traditional economic models and adopting more inclusive and sustainable approaches to growth and development. Social challenges like poverty, inequality, and social exclusion are central to the sustainable development agenda. Ensuring that all individuals and communities benefit from development efforts is essential for achieving a just and equitable world. Improving the root causes of poverty and inequality and promoting inclusive development through education, healthcare, and social protection are crucial for achieving equitable progress.

Sustainable development's institutional and policy challenges highlight the importance of effective governance, policy coherence, and international cooperation. Strengthening global governance structures, building institutional capacity and enhancing policy coherence are critical for translating global goals into actionable strategies and practices. Evaluation metrics are crucial in tracking progress and ensuring accountability in sustainable development efforts. Developing more comprehensive, context-specific, and participatory approaches to evaluation can help ensure that sustainability is measured accurately, inclusively and relevantly. As we look to the future, the sustainable development agenda will be shaped by emerging trends and issues, such as technological innovation, systemic change, innovative financing and social and cultural shifts. Navigating these challenges and opportunities will require strong leadership, global cooperation, and a commitment to equity and justice.

In sum, the journey towards sustainable development is a collective endeavour involving all societal sectors, from governments and businesses to civil society and individuals. While the challenges are significant, the potential for positive change is immense. By working together in a spirit of solidarity and cooperation, the global community can overcome the obstacles to sustainability and create a better future for all.

Index

www.ingramcontent.com/pod-product-compliance
Ingram Content Group UK Ltd.
Pitfield, Milton Keynes, MK11 3LW, UK
UKHW021852100926
8556IPUK00020B/1194

9 781803 746616